Waste Management in Jasper National Park

Waste Management in Jasper National Park

by

Katherine M. Calvert,
Richard D. Revel and Leonard V. Hills

Parks Canada

Canadian Cataloguing in Publication Data

Calvert, Katherine M., 1947-
Waste management in Jasper National Park

ISBN 0-662-23059-0

1. Refuse and refuse disposal—Alberta—Jasper National Park. 2. Jasper National Park (Alta.)—Management. I. Revel, Richard David, 1946- II. Hills, L.V. III. Parks Canada. IV. Title.

TD789.C32J37 1995 363.72'8'09712332 C95-910320-1

Printed on acid-free paper.

Typesetting: By Design Services, Calgary, Alberta, Canada

Printed and bound in Canada

Contents

List of Figures

List of Tables

Preface

Social, environmental and economic concerns about waste management are increasing as populations increase and society becomes aware of both the cumulative and long-term impacts of the disposal of our waste products.

Much research has been undertaken and greatly improved waste management practices have been implemented in large urban centres. Less fundamental research, however, has been forthcoming from smaller communities, rural areas or large parks. Due to the exceptionally time-consuming, tedious and expensive nature of waste stream analysis, little quantitative data have been collected to give guidance to waste managers on the nature of the product they are managing and most of this information relates to urban environments. No comprehensive information was found pertaining specifically to large parks.

Jasper National Park, established in 1907, is a very large park (10,800 square kilometres), the majority of which is pristine and includes immensely varied terrain from grassland through forest to alpine. The park is quite unique in that it includes the town of Jasper within its boundaries. Jasper has a permanent population of 4,500 year-round which is augmented by as much as an additional 1,500 people daily during the tourist season.

Because of the 'town within the park' context of Jasper National Park, this park is an ideal study site for detailed investigations on waste management as it addresses both high tourism and park-use issues as well as waste stream attributes of smaller urban towns. Our study addresses the overall issues of waste management in the park as well as providing detailed and comprehensive data on the composition of the waste stream in both the park and the town on a sector by sector basis as well as on a seasonal basis.

We have organised the document in a report style by providing the reader with background on both the park and its present waste management as well as a brief overview of waste management in the North American context. We then move on to a review of waste management in other national parks and small towns. Having once set the stage, a methodology is set out for our Jasper study followed by a very comprehensive, and seasonally organised, data presentation on the waste streams from commercial, residential and campground sources. This detailed waste stream composition analysis is then used as the basis for an evaluation of waste management, transportation, economic and recycling options for the park.

We anticipate that our investigations will be of considerable interest to all those charged with waste managementportfolios in large or small parks or other tourist-driven facilities as well as small towns in rural settings. While not particularly directed at large urban communities, undoubtedly the comprehensive waste stream analysis will be of considerable interest in that context.

Katherine Calvert, Richard Revel, and Leonard Hills

The University of Calgary, March, 1995

Acknowledgements

We owe a debt of gratitude to many institutions who have provided financial support, technical expertise or preparatory support during the development of this book.

To Department of Canadian Heritage, Parks Canada and specifically Jasper National Park a great deal of credit must be recognized for both technical and financial support. Jasper National Park sponsored and financed the project. Many people within the parks system provided immeasurable help in surmounting the numerous problems which crept up at various stages. Gaby Fortin, former Superintendent of JNP provided unrelenting commitment to the work and did everything possible to make the research run smoothly, Maryse Blouin and Mark Kolasinski provided excellent computer skills and editorial input when it came to producing the final disks, truck drivers often arranged their trucking schedules to accommodate the research needs and personnel at the transfer station assisted with the work despite inconveniences. Andy Walker was instrumental in seeing the publication to its conclusion and must be thanked for his support.

Kevin Metcalf and Norm Nuttal made their facilities available for research during the early stage of the work particularly in the area of statistical sampling. Lelani Arris provided volunteer assistance with statistical analysis throughout the work making this project a reality.

The senior author received an M.Sc. in the Resources and Environment graduate program at the University of Calgary under the supervision of the book's junior authors. That work constituted the basis of the present book.

The interdisciplinary foundations of the Resources and the Environment program allowed for the involvement of individuals with varying academic expertise which would not have been possible were the work undertaken in a conventional core faculty. These individuals, both within and without the university are too numerous to list in the acknowledgements, their names and connections are listed throughout the text as personal communications. To these individuals we owe a deep debt of gratitude.

The input and advice of many individuals went into the final production of this book. That input and advice is much appreciated, the strengths of the work must be shared with them. The ultimate responsibility for any weaknesses remains ours.

Katherine Calvert, Richard Revel, and Leonard Hills

The University of Calgary, March, 1995

Dedication

Toward improved waste management in Canada's national parks and specifically in Jasper National Park.

1 Introduction

JASPER: THE PARK AND TOWNSITE

Jasper National Park was established in 1907, shortly after Banff National Park. In 1910, the Grand Trunk Pacific Railway selected the present location of Jasper as its divisional point. The town prospered under the influence of both the railroad and the tourist industry and presently supports a permanent population of approximately 4,475 people (pers. comm., Chambers, 1993).

Although the park is a well known tourist destination, unlike Banff, the town remains somewhat insulated from the effects of tourism. The major tourist influx in Jasper is in July and August, whereas, Banff is subject to heavy tourist traffic all year. Much of this is due to the park and townsite location.

Jasper is located 386 km west of Edmonton, 427 km northwest of Calgary and 824 km northeast of Vancouver. Connecting highways are #16 from Edmonton and #93 from Lake Louise. Although the route to Vancouver (termed the Yellowhead) is well developed, it does not account for the majority of east to west traffic. Isolation is popular with the permanent residents and at this point, unlike Banff residents, they express no desire for autonomy from the park.

The park currently supports 11 hotels, 10 apartment buildings, 193 condominiums, 160 commercial lots, 19 institutions, 95 business properties (including the industrial park) and approximately 900 residences. There are 9 motels, 50 summer residences and 10 campgrounds outside the townsite. Tourism is a major industry with 2.5 million people visiting each year. In August on a peak day the number of visitors in the park can be as high as 15,000 people (per. comm., Breau, June 1991).

The type of waste dominating the commercial sector reflects the type of garbage generated by motels, bungalows, restaurants and other service type businesses. This waste is the result of the economy of the park being strongly centred in the tourist industry.

Jasper is a unique community in that the town is located in a National Park and the predominant commerce is based on the tourist industry (pers. comm., Audy, 1993). To establish an effective waste management strategy for the park it is necessary to first determine the nature of the waste stream (Howelette, 1990) produced in such a tourist centred economy. This includes establishing the composition of the waste, its principle source, and the amount produced. This information can then be used to determine the options available to the park to reduce the amount of waste generated in the park. These options include reducing the source of waste or recycling usable material introduced into the waste stream.

GOALS

Every community has its own recycling and waste reduction potential. Depending on local conditions, which include the characteristics of the waste stream, development constraints, level of waste flow control, environmental regulations, financial capabilities, material markets, public and political acceptance, it is possible to assess the level at which the waste may be produced (O'Brian and William, 1991).

The incentive to do a survey of the solid and trade waste produced in Jasper stemmed from the fact it is a National Park with a large seasonal variation in visitation. Because of the strong influence tourism has on the park and the business it generates, it was believed that the waste stream would differ from that of other urban or rural Alberta communities. It would be unrealistic to take waste analysis data from a typical Canadian city and apply it to Jasper with any degree of accuracy.

In this regard, significant variations and differences exist between the refuse produced in different cities and different rural communities, and it is reasonable to assume that Jasper will have its own problems (Van Den Broek and Kiror, et al., 1969, Lilley, 1985). The ultimate intent of this research is to establish base line data from which a realistic recycling program can be developed to meet the 50% waste reduction goal targeted in the federal government's Green Plan for the year 2000 (*A Framework for Discussion on the Environment, The Green Plan*, 1990).

An additional goal is to establish the importance of transportation in removing waste from the park. The major cost of garbage disposal comes through collection and removal of this material. Because this is an important part of the parks occupation in waste disposal, a chapter has been devoted to this issue.

OBJECTIVES

In setting out to examine waste management in Jasper National Park, we established the following objectives:

1) Determine the nature and source of the predominant waste stream in Jasper National Park;
2) Examine ways to reduce the waste, by determining what material can be eliminated through reduction or recycling;
3) Examine the transportation costs to export garbage;
4) Examine the present recycling efforts in the community and the extent to which these efforts reduce the waste stream;
5) Examine the recycling activities of communities similar to Jasper and determine if their programs are applicable in this community.

PRESENTATION OF MATERIAL

The majority of presentation centres around the results of an analysis of the waste that passes through the Jasper transfer station as this is the handling location of most day to day material. This information will then be used to assess how the waste can either be reduced or recycled given the resources of the park.

Particular elements that make Jasper a unique community are presented, and recommendations for a specific waste reduction program may be established. An assessment of waste stream management in other National parks and communities in Western Canada is taken into consideration to determine which programs, if any, are applicable to Jasper.

In this assessment, transportation will be considered an important factor in how the waste is disposed of.

DEFINITION OF TERMS

CPS: Canadian Parks Service

Generators: The true sources of garbage sampled, these being commercial waste, residential waste and campground waste. A fourth type was mixed garbage from outlying areas that is a combination of the other three types.

JNP: Jasper National Park

MSW: Municipal Solid Waste.

Recycled Goods: That material which is discarded and is part of the waste stream, but which has a value as a commodity and is diverted from the municipal solid waste through the recycling process.

Sanitary Landfill: This is a landfill site that accepts MSW or products discarded by commercial or residential units. Jasper currently does not have a sanitary landfill as all material is taken to Hinton.

Transfer Station: A building constructed to hold MSW collected from the park to be transferred to the Hinton landfill for permanent disposal.

Trade Waste: Waste that is not considered municipal solid waste such as construction debris, old fill, cars, tires and lumber. This material is commonly landfilled at a trade waste pit.

Trade waste pit: Land set aside to accept trade waste material.

White Goods: Standard household appliances such as refrigerators and stoves.

READING THE GRAPHS

Because many of the materials sorted in this study constitute less than 5% of the waste stream, they are put into one category which is labelled 'other'.

2 Background

HISTORY OF WASTE MANAGEMENT IN JASPER NATIONAL PARK

Old Landfill Sites

There are five landfill sites in JNP (Figure 1) one of which is the trade waste pit in current use. During the summer of 1992 the parks selected three sites to be inspected for contamination. These sites are the:

1) current trade waste pit.
2) former dump at the Jasper wood lot.
3) former dump and incineration site on Pyramid Bench.

Each site was graded according to the standards set by the National Contaminated Sites Remediation Program (*National Contaminated Sites Remediation Program, National Classification System for Contaminated Sites,* 1991). This grading provides the only recorded information kept on old landfill sites in Jasper. All other information on these areas comes from personal communication with those who worked in the area of garbage disposal at that time.

Prior to the use of the current sanitary landfill and the trade waste pit, all waste, regardless of source, was dumped into what is now called the wood lot, an area on the south side of town that is now used to store wood for the campgrounds. When contamination of the Miette River became unacceptable, this location was changed to a ravine one half mile north of the townsite on the Pyramid Bench road (figure 1) (pers. comm., Stendie, 1993). By 1969, the amount of garbage in the ravine had reached Cotton Creek. This lead the park to believe a serious leachate problem could develop into this water system (pers. comm., Edwards, 1991). The water has not been tested, but observations of old landfill sites not otherwise tested for leachate problems suggests that they may all have this problem (O'Leary and Walsh, 1991).

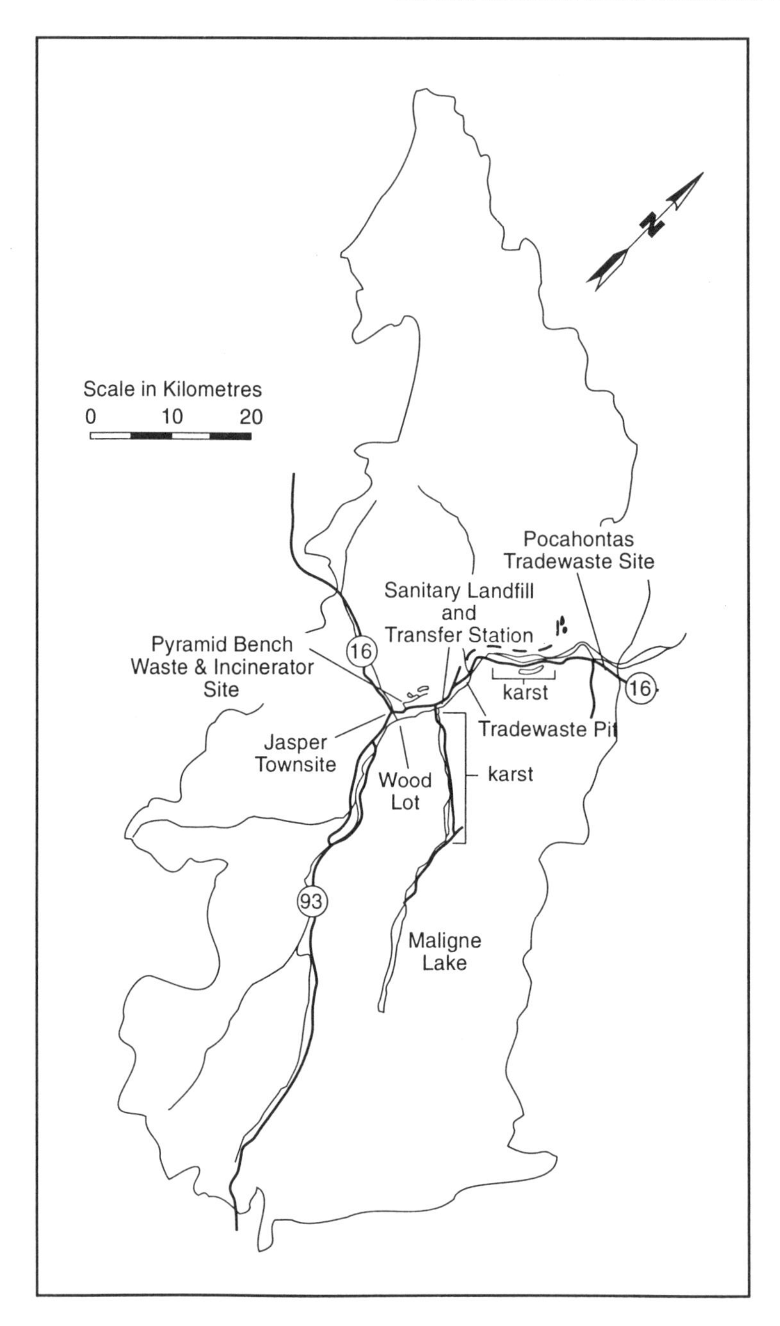

Figure 1 Jasper National Park Waste Pit Locations

This site also housed a brick incinerator that burned garbage brought in by the Park Service. The incinerator was very old and not capable of burning refuse to the emission standards designated by the Province of Alberta. It was dismantled when a new location for the landfill was acquired (pers. comm., Don Lonsberry).

Other sites included a trade waste pit at Pocahontas, 56 kilometres east of Jasper on the Miette Road and the Maligne Lake pit, located on the skyline trail, one kilometre from the lodge area. The Pocahontas site serviced the Pocahontas campground, warden station and bungalows. This site was closed when bear proof garbage bins were installed and waste was trucked to the sanitary landfill (Figure 1).

Both of these sites were covered with local fill but were not capped according to Provincial Standards, nor was any attempt made to contain leachates (*Code of Good Practice on Dump Closing*, 1977).

By 1969, both the Pyramid and Pocahontas landfills had reached maximum capacity and a new site was required. Neither of the above sites were enclosed and therefore posed a hazard to the public from frequent habitation by bears. Prior to this time, such dumps were a tourist attraction for the public to observe bears. This lead to the inevitable bear/human conflict. To alleviate this problem it was decided that bears must be excluded from dumps, an abnormal food source that leads to undue familiarization with people and consequently bear-human confrontations.

The decision to relocate the dump site prompted the park to assess the type of garbage disposed by the public and the government (pers. comm., Edwards, 1992). This review lead to the creation of two new sites: the sanitary landfill, which received municipal solid waste, and the trade waste pit, where dry waste such as construction debris, large household goods and yard waste were disposed.

The Sanitary Landfill

The current sanitary landfill was established in 1969, but was not in full use until 1970. It is located approximately 10 km. east of Jasper townsite, on high ground, to the north of the highway. Formerly an old borrow pit (gravel extraction site), the open ground was considered suitable for a sanitary landfill site. This site received all solid waste from the park, as well as animal carcasses disposed of by the warden service.

The sanitary landfill was used in this capacity until the establishment of the present transfer station. Two wells were established to test the leachate emission from the former dump site. The up-gradiant well was established above the location of the present transfer station and has tested negative for any leachates or contamination. This water is currently used as a water source for the station's requirements.

A down gradient well near the highway below the dump, has never been tested. There is some concern that the well head may be too close to the dump to detect any significant leachate problems. The site was capped with soil located within the vicinity of the dump, and subsequently rehabilitated. A clay cap was not used, which may lead to problems with water contamination in the future (*National Guides for Landfills of Hazardous Wastes*, 1991). At this point there is no noticeable settlement; however the cap has only been in place for one year. Problems that have occurred at this site are development of leachates and methane gas (pers. comm., Stendie, 1979). The site is now used to store materials which can be recycled but previously were dumped at the Trade Waste Pit.

The Trade Waste Pit

The Trade Waste Pit, located 13 km. east of Jasper, was originally a borrow pit for gravel and sand used in highway construction. It became the new location for trade waste in 1969, when the Pyramid Bench disposal site was closed. This location for the Trade Waste Pit was convenient at the time and originally the waste was deposited on a section of land just north and west of the present landfill (pers. comm., D. Lonsberry, 1990) (Figure 2).

As the pit reached capacity, the waste was deposited adjacent to this spot in a corner to the southwest. This brought the dump closer to the Athabasca River, where the water table is near the surface. As a result, trade waste is now being dumped into surface water. There is no known containment of leachates and the pit is on an alluvial fan (pers. comm., Marsh, 1992). The aquifer layer is 28.7 m below the landfill and continues to at least 54 m. This information was obtained from the logs of two water wells located 1.5 km upstream of the pit (*Mike Knauer Report, 1986*).

The Trade Waste Pit was originally monitored to prevent scavenging and unauthorized dumping of waste materials not designated for this site

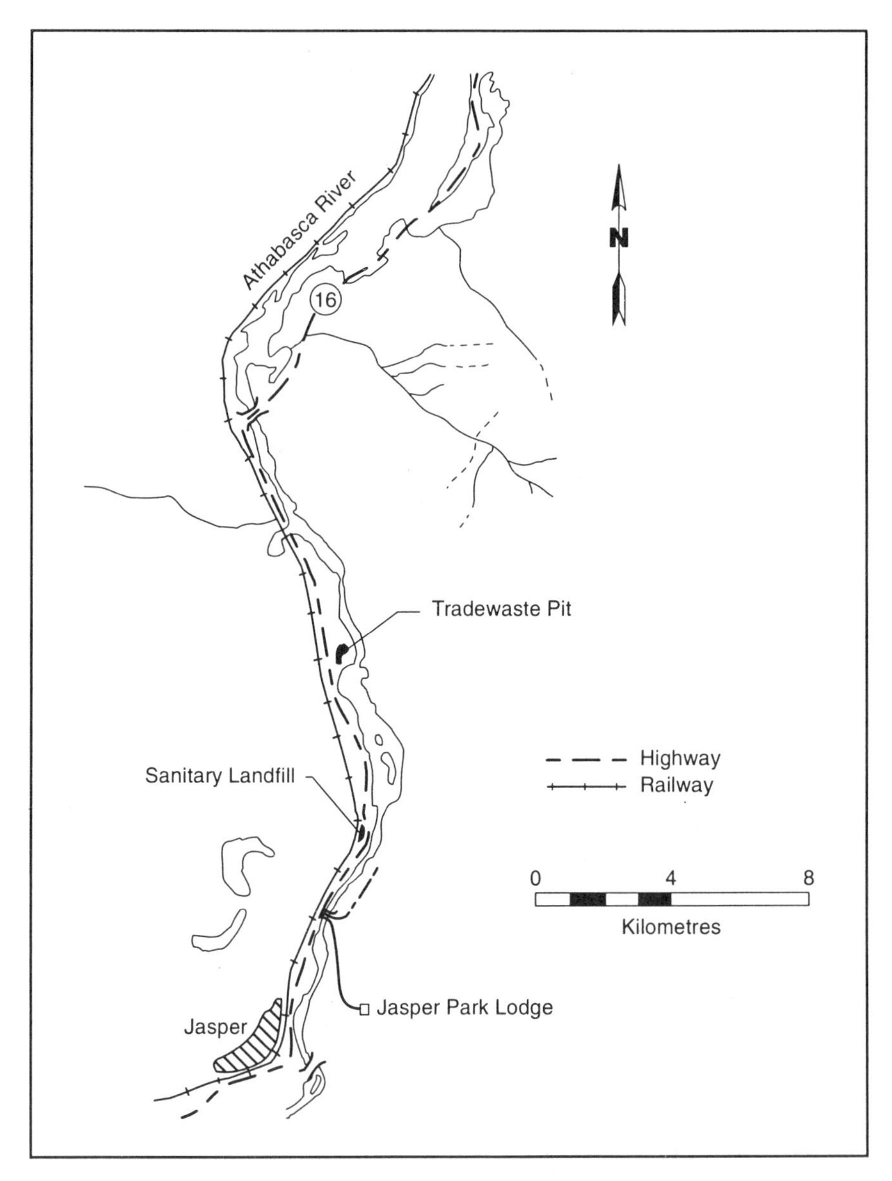

Figure 2 The Tradewaste Pit And The Sanitary Landfill Locations

(pers. comm., Dave Edwards, 1990). A gate was erected at the entrance and locked after dump hours. This proved to be unpopular with the local population, who salvaged material from the dump site. This lead to the gate being cut down allowing for repeated night access to the dump (pers. comm., Edwards, 1990). Authority over the site alternated between General Works and the Warden Service, with neither group establishing any real control. As a result there has been indiscriminate dumping for several years.

Early in the 1980s, the CNR began a program of switching the wood ties, used to support the track, to cement ties. The removed ties, which are preserved with creosote, were buried in the trade waste pit. The number of ties can only be approximated by parks and CNR officials (pers. comm., Lonsberry, 1992). Other contaminants have been estimated by the park to consist of goods listed in Table 1, but the quantities deposited are not known. There have been no wells established on this site; therefore the problems with leachates are unknown. This site, along with the wood lot site, have been recommended for further investigation regarding the amount of pollutants that may be entering the water system. Consultants have been approached to provide an assessment of what would be required to clean up the potential pollution problem; however, this will depend on the findings of the initial survey (pers. comm., Marsh, 1993).

Finding contamination may be hampered by downward drainage of water into karst aquifers that may connect with the Rocky and Maligne river systems. A karst formation is any location in the landscape which is being altered markedly by solution, generally resulting in a cave system. Karst formations tend to develop in any of six limestone rock units of the Rocky Mountains, one of which lies under or near the trade waste pit (Gadd, 1986).

At the current rate of filling, the Trade Waste Pit will be full in the near future and a new site will have to be established. Money is now being sought to build in a weighing and monitoring facility for the new location (pers. comm., A. Walker, 1993).

Current Collection Methods

All solid waste in Jasper is deposited in green, bear proof bins. The garbage is collected with trucks that have a hydraulic lift to raise the bins and allow the contents to be deposited into the holding container. Once the garbage is in the truck it is compressed to compensate for

Table 1 Jasper National Park Inventory of Contaminated Sites

A list of materials observed at the trade waste pit site includes:

- Tires
- Batteries
- Car Bodies (engines, gas tanks)
- Building Demolition Rubble (brick, concrete, wood, plaster, insulation, etc.,)
- Metal
- Plastic Containers (oil, pesticide, antifreeze)
- 45 Gallon Drums, 5 Gallon pails, Paint Cans
- Wood and Wood Products
- Paper, Glass, Plastics
- Tree Branches, Clippings
- White Goods (refridgerators, washing machines)
- Carpeting, Mattresses, Etc.,

List of possible hazardous chemicals:

- Solvents, Paints, Lacquers, Stains, Etc.,
- Resins
- Wood Preservatives
- Petrochemical Products
- Heavy Metals - lead, mercury, chromium, etc.,
- Acid, Bases
- Photodeveloping Chemicals

bulky objects such as cardboard. The garbage is then taken to the transfer station, where it is dumped and removed to Hinton in a transfer trailer (pers. comm., Ross Pigeon, 1990). The maximum allowable weight carried by the truck is limited by the highway restrictions imposed by the province. The collection problems will be addressed in Chapter 6, Transportation.

Garbage Sources Within the Park

Park garbage can be divided into four categories: commercial waste, residential waste, campground waste, and roadside or mixed waste from outlying areas.

Commercial Waste: Commercial waste is generated from the hotels, restaurants, bungalows and food outlets such as IGA, Super A and the small retail outlets, Canadian National Railway, all businesses in the industrial section of Jasper, the post office, hospital and schools. Commercial pickup is separate from the residential pickup and is a pure source of this waste material.

Residential Waste: Residential waste is picked up in two or three specific runs each week day, with the exception of Saturday. Residential waste is generated by permanent park residents living in single family units or multi-dwelling apartment buildings and is not influenced by the tourist industry.

Campground Waste: Campground waste is seasonal and does not develop significant amounts until June, with the start of the summer tourist season. The bulk of this waste is generated throughout July and August, and tapers off after the long weekend in September ceasing by the end of the month. There was only enough waste generated from this source to collect 1 load a day. A second truck was used to pick up ashes from fire pits during the busy part of the season.

Mixed Waste: Mixed waste is garbage from bungalows, campgrounds and roadside pickups and is comprised of mixed garbage that cannot be categorized into one single source. This waste comes from all the outlying areas of the park. The collection runs are on the two highways that serve the park, plus the backcountry areas of Maligne Lake, Miette Hotsprings and Edith Cavell. The garbage that is picked up is a mixture of bungalow, campground and roadside pick up. During the high point of the summer this amounts to 30% of the daily garbage collected.

THE JASPER RECYCLING SOCIETY

History and Role in the Community

The Jasper Recycling Society was founded in the fall of 1989, as a non-profit group to promote the reduction of Park waste. The original committee was large and the first meetings concentrated on establishing what and how to recycle marketable goods. Establishing a blue box program was the main objective of the program which concentrated on recycling tin, aluminum, paper and glass. In conjunction with the blue box program an educational program was directed towards the business and commercial sector of the community. The information reached the town residents through the local newspaper and monthly newsletters.

In 1989, the society received a grant from Alberta Environment and purchased 400 blue boxes and eight green collection bins for glass, aluminum and tin. The bins were placed in the Industrial Park, south of the Jasper townsite. The boxes were distributed to local households, hotels, and businesses in Jasper.

This was followed by their undertaking a variety of successful programs which include:

1) Providing local grocery stores with clean, used bags for the customers. The bags were purchased with funds raised by the recycling society and donated to the stores;

2) Convincing the local retailers to support recycling by putting up posters on recycling and other environmental issues;

3) Educating retailers to bring in recycled and environmentally friendly goods, such as products with less packaging. This project is not yet complete;

4) Promoting recycled paper through advertising the paper recycling facility provided by the senior citizens at the museum parking lot;

5) Publishing articles on the Jasper Recycling Society and recycling in the local newspaper. This included announcements of the societies, meetings plus an update on the amount of material being recycled through the blue box and green bin program;

6) Attempting to establish a blue box in the post office. However, this was rejected by the Supervisors from Canada Post. A poster provided by the Jasper Recycling Society has been placed in the post office, with regard to the rejection of the blue box. The Society is continuing in its efforts to have Canada Post change their position; and

7) Participating in the annual toxic waste round-up organized by the Canadian Park Service. Society members help to collect waste dropped off at the collection site. They are also available to provide any information on their current recycling program and to promote consumer awareness.

Current Work and Projected Goals

The ultimate goal of the Recycling Society is to involve the Canadian Parks Service, along with other outside agencies such as the bottle depot, in recycling endeavors (pers. comm., Landry, 1991). Although community support is necessary to maintain a viable recycling program, much of this work can be handled by establishing a permanent program, that allows for maximum recycling and waste reduction. The recycling society, with the support of the government, is now focused on recycling cardboard. The current goal is to reduce the amount of cardboard going to the landfill by over 50% in the next year (pers. comm., Landry, Sept, 1992).

PRESENT RECYCLING IN JASPER

At the beginning of this study, there was little waste recycled in the park. The Jasper Recycling Society was newly formed and the town had no bottle depot. All glass and aluminum containers were returned in person, to the recycling depot in Hinton, or picked up through occasional bottle roundups sponsored by local community groups. The amount of material recycled was estimated to be around 1% of the material landfilled through the Transfer station (per. comm., Edwards, 1991).

Presently, paper is being recycled by the senior citizens at a trailer located on the museum parking lot in town. This is labour intensive work that is limited by a small facility.

In 1991, Jasper finally got a bottle depot, which quickly accounted for a much greater proportion of the glass, tin, aluminum and plastic bottles being recycled. Also, at this time, a larger recycling depot was built at Whistlers Campground, replacing a small, poorly located shed previously used. The campground is large and complicated, making it difficult to find the depot.

The Jasper Bottle Depot

Information on the total amount of glass, metal, plastic and aluminum was obtained from the depot for the months of July, 1991, through to the end of March, 1992. Since all this material was recorded at the depot by volume, each category was weighed according to the flat size used for shipping. This weight was then calculated by the numbers of each unit shipped per month to gain a monthly total by weight. As each type of container has a different economic value there were several different categories for each material. Since there was no sampling during October, November and December, 1991, these months were not included. The September values were taken to replace the lack of data for the month of June, when sorting first began at the transfer station.

The truck weights were totalled for the months between June and March, excluding the 3 fall months. The average percent of each category was calculated over these months and multiplied by the total weight to give the individual weights for each category over this period of time. The depot weights were divided by the total and multiplied by 100 to give the percent each category was being recycled out of the waste system.

Both glass and aluminum appear to be a small part of the waste stream as indicated from their low percentage rate in overall waste composition (figure 19) at the transfer station. However, the number of containers used in our consumer society suggests they should show up some where in the waste stream. It was gratifying to see that most of the glass and aluminum can be accounted for. Plastic was a different problem. The depot only takes plastic in the form of beverage containers, however, the plastic separated at the transfer station consisted mostly of packaging material. Undoubtedly, most of the plastic discarded in Jasper is going through the transfer station, but, it is so light it is still not a significant percentage in the waste stream. The metal separated at the transfer station is often in a form other than metal cans which is all that is accepted by the bottle depot. Much of this waste appears to be going to the trade waste pit.

Senior Citizens Paper Recycling

The same approach was taken as for glass etc, to determine the effectiveness of the senior citizens' recycling efforts. They have kept all of their records for the last several years so information was obtainable for each type of separated paper and the amount of cardboard they received. One problem, however, is that the loads are trucked off infrequently and, until a whole year of data is in, it is difficult to match their recorded weights with the percent composition over this period of time. The data available at this point was from July 22, 1991, to May 27, 1992. All truck weights from the transfer station were available for these months, however, the compositional analysis was only complete from January to May. The average amount of paper in the waste stream was 15.6% and it was assumed that the inclusion of December and

Table 2 Bottle Depot Total Weights

July, 1991 to March, 1992

Glass	Plastic	Aluminium	Metal
174,397 kg	4,420 kg	71,887 kg	133 kg

Transfer Station Total Truck Weight

June, July, August, January, February, March

1,868,178 Kilograms

Average Percent Composition and Weight

Glass	Plastic	Aluminium	Metal
4.25%	7.6%	1.25%	3.8%
79,398 kg	141,982 kg	23,352 kg	70,190 kg

Percent of Material Recovered

Plastic	4,420 ÷ 146,402 × 100	= 3.0%
Aluminium	71,887 ÷ 95,239 × 100	= 75.5%
Metal	133 ÷ 70,323 × 100	= 0.19%
Glass	174,397 ÷ 253,795 × 100	= 68.7%

November in the fall of 1992 would not alter this significantly.

The total amount of paper collected was 131,500 kg. This included newsprint, computer paper and coloured paper. Glossy magazines were rejected, as were glossy flyers.

The total amount of paper weighed at the Transfer Station during this period was 457,725 kg. The percent of recycled paper is the total of recycled paper divided by the sum total of recycled paper and waste paper, multiplied by 100.

Percent Recycled paper: 131,500 + 457,725 = 589,225
131,500 ÷ 589,225 × 100 = 22.3%

Thus the senior citizens account for 22.3% of paper removed from the waste stream. Since they opened this centre in 1989, they have recycled 96 tonnes of paper (as of 1990).

Future Recycling in Jasper

The immediate goal for the park is to establish cardboard recycling in the spring of 1993. Some problems are still not solved, such as how to get the cardboard to the recycling depot, or how to store it. Transport options have been investigated with various trucking companies; however, the best option to date is rail transport. The cardboard will be shipped to Paperboard in Burnaby, BC. The Park estimates on the basis of last summer's study, that 540 tonnes of cardboard will be available for recycling in one year. Rail transport can take 75 tonnes in one rail car per trip. The distance is 960 km. and the cost per trip is $1,000. This gives a cost of 1.4 cents per tonne per km. This does not include the truck costs to haul cardboard to the rail head. There are no trucking transport costs in Vancouver as the rail line goes directly to the recycling station. Once the depot is running at full capacity, paper will also be recyled through this facility.

CURRENT WASTE MANAGEMENT IN JASPER NATIONAL PARK

The Jasper Transfer Station

The Jasper Transfer Station was developed in response to a series of meetings regarding the formation of a Regional Solid Waste Management Authority. This Authority would manage and control the efficient disposal of all solid waste generated within the Yellowhead corridor between Jasper and Edson. These meetings were held in 1979 and early 1980 between Alberta Environment, the Alberta West Central Health Unit and representatives of the Canadian Park Service (pers. comm., Edwards, 1992). Dillon Consulting Engineers were commissioned, by Alberta Environment, to prepare a solid waste management study for the corridor. Dillon and Assoc., 1981, recommended that the most efficient way to handle the waste produced in the Yellowhead District was to develop a regional landfill in the Hinton area. The community of Edson, and smaller communities in Improvement District 14, would develop transfer stations to collect local waste and truck it to the regional landfill. Jasper was included in this proposal and subsequently installed the transfer station for collection and disposal of waste at the Hinton Regional landfill.

The transfer station in Jasper was completed in March, 1990 and has handled all municipal solid waste in the park since it's inception. The station is located 5 kilometres east of the Jasper townsite and receives municipal waste from the four sources listed in Chapter 2. The transfer station is located on the site of the old sanitary landfill which previously received all municipal solid waste. The site is enclosed with an electrified wire mesh fence that repels large mammals such as bears, elk and coyotes.

Recycling Units at the Transfer Station

Five cement recycling bins were built at the transfer station during the installation of the building, with money from the Park and donations from the Jasper Recycling Society. The recyclable material comes from the eight green bins in Jasper that receive glass, tin and aluminum. When a full trailer load has been collected, the material is taken to Edmonton for recycling. The Recycling Society hoped to purchase a glass crusher to reduce the volume of the glass transferred to Edmonton, but was not able to do so with the funds available. At present, the glass is shipped as received. The sanitary landfill is the storage area for other recyclable material such as tires, white goods, asphalt and wood products. All of this material has accumulted on site and has not been recycled at the time of writing.

The Hinton Regional Landfill

In 1981, Alberta Environment requested a waste management study from Dillon and Associates for the Edson-Hinton corridor, which also included Jasper and Switzer Provincial Park (Dillon and Assoc., 1981). The primary objective was to develop a regional solid waste management plan that would ensure that waste disposal was handled in an economic and environmentally sound manner. The specified time given for this disposal was from the year 1981 to the year 2000.

The Solid Waste Management Assistance Program was established to help municipalities in Alberta, but, the program excluded federal establishments such as Jasper National Park. However, in view of the proximity and interrelationship with the park to the area, the terms of reference allowed for the inclusion of Jasper municipal waste. This study resulted in the development of the Regional Landfill at Hinton to accommodate both the local and park waste. A transfer station was built in Jasper. During initial negotiations trade waste was also included, but, has since been excluded. Jasper is now resposible for the disposal of trade waste within the park.

3 Waste Management in North America

A REVIEW OF WASTE MANAGEMENT OPTIONS

The direct disposal of waste is handled mainly by landfilling or incineration. Other alternatives for hazardous waste are deep sea dumping, incineration, or deep well injection (pers. comm., Boyle, 1992). Because of the costs associated with waste management and the potential harm to the environment, there must be a re-evaluation of the current practices (*Recycling of Waste in Alberta, Environment Council of Alberta,* 1987). This does not mean that landfilling or incineration are improper disposal methods; however, the alternatives of reduction and recycling must be considered to reduce the amount of waste going into these facilities (Howlett, 1990). Waste does not become waste until it enters the waste stream. Therefore, it can be said that there is potential waste and actual waste. Waste can be diverted from the waste stream by not generating the waste in the first place (reduction) or diverting it by reuse. Once the disposed article has entered the waste stream it may be recycled or composted. Composting, in this regard, is recycling of compostable wastes.

Landfills

Landfilling is currently the cheapest and most common method of waste disposal (Leone, et al., 1977). Landfills vary from simple disposal of dry uncontaminated waste (trade waste pits) to the disposal of municipal solid waste contaminated with hazardous products (*Guidelines for Industrial Landfills,* 1987). Poor handling of waste in unacceptable terrain has given the landfill a bad image, making the relocation of such sites very difficult. There are concerns of ground water pollution, litter, odors and gas elimination, as well as increased heavy truck traffic (Devries and Ross, 1990). Poor management practices of landfills in the past have left people skeptical of the ability to develop properly run facilities that can be rehabilitated. Consequently, property located close to a landfill generally suffers from decreased land value (*East Calgary Regional Concept Plan,* 1989; Edwards, 1981).

Despite the poor image now associated with landfills, they will remain the chief resolution to garbage disposal until such time as waste reduction and recycling can counteract the increasing volume of waste produced. Under current engineering practices landfills can be sanitary and fully reclaimable as useful land. The advantages of a well planned landfill have been cited by the Institute for Solid Waste of American Public Works Association. A well developed landfill can reclaim useless land and provide a disposal site for all classes of solid waste without segregation (Howlett, 1990). Several communities in the United States have rehabilitated landfills to the point of building golf courses, wildlife habitat and even housing expansions (Ettala, 1991).

In order to allow for the successful rehabilitation of this land, the initial design must include a proper lining of the landfill, preferably with a dense clay and a plastic liner. The final capping is critical to prevent water infiltration into the landfill. Water is responsible for the development of leachates and methane gas and acts as a catalyst for chemical reactions of the dumped material (O'Leary and Tansel, 1986). Although initially costly, with proper handling and compaction of the dumped material, lining and capping the landfill according to the code specified by Alberta Environment, landfills can be acceptable for waste deposition and still remain the cheapest alternative for waste management.

Incineration

Incineration is the most commonly used alternative to landfilling (pers. comm., Carol Boyle, 1992) and as land becomes unavailable this may become a more desirable alternative. The ash of the incinerated waste is less bulky and is more easily handled than solid waste (Webb, 1983). However, capital and operating costs are high if the emission level is to remain below the standards set by Environment Canada. Garbage incineration is thought to be the largest single source of dioxin production in Canada (Rahn, 1987).

Reduction

The prime goal of a waste reduction program is to not produce the waste in the first place. It is a preventive measure that eliminates all of the other steps needed to reduce the waste that comes potentially from any product or material produced for our consumption. In a consumer society not producing products in the first place, is considered by some to be contrary to our concept of economic growth. It may result in a lower standard of living if the standard of living is defined by the amount of material consumed (Freeman, 1973).

Another form of waste reduction is to reuse potential waste before it becomes part of the waste stream. Reuse of materials usually refers to the repeated use of a product in the same form. The product can be used in its intended form or for another purpose (Howlett, 1990). This has the same effect on the economy as recycling in that new goods are not produced, therefore, manufacturers lose business. There is a benefit to the consumer, however, from either waste reduction or reuse, in that fewer purchases are needed. To implement either action, a fundamental restructuring of our thinking will be required (Hays, 1978).

Recycling

Recycling is the most popular and environmentally acceptable way to dispose of refuse (Goldin, 1987). Recycling is a form of reuse as the product goes back into the environment it came from or is redirected to another product. The current interest in recycling comes from our concern for the environment, along with concern for our standard of living. The expansion of landfills, particularly the massive inland landfills being developed in the United States, indicates there is a threat to the environment, living conditions and the economy (Woods, 1991).

Composting

Successful recycling programs are based on a realistic assessment of the waste stream, and composting must be seen as part of the recycling program. Composting has been defined as:

> The controlled (aerobic) biological decomposition of organic material to a point where storage, handling and land application can be achieved without adversely affecting the environment (Hoitink, 1990, in Burnett, 1991).

Composting is the oldest and most universally practiced form of soil treatment in the world (Burnett, 1991). There are some basic principles which must be attended to, however, to ensure a successful compost (Hoitink, 1990). First, a useful community composting system must be viewed from the viability of their product and the ultimate use it will be put to. The turnover time of the compost must also be a community consideration. The longer it takes to create a viable compost, the more expensive that product becomes.

WASTE MANAGEMENT IN NATIONAL PARKS AND RELATED COMMUNITIES

An analysis of waste programs in all Canadian National Parks and related communities is not the purpose of this study. Although there are many common denominators in waste management, the regional geographical differences create specific problems for specific areas. Therefore the parks and communities reviewed will be confined to the Canadian Rockies. This region includes the Four Mountain Parks, Waterton Lakes National Park, the Bow and Yellowhead Corridors and small communities where specific management programs are being developed.

National Parks Solid Waste Policy and Legislation

The Federal Green Plan, 1990, proposed the creation of an Office of Waste Management to promote waste reduction and recycling. It sets a goal of a 50% reduction in waste which we should attain by the year 2000.

The Park Policy and Procedures Manual, (1980), dictates that natural resources within National Parks will be given the highest degree of protection to ensure the perpetuation of the natural environment, essentially unaltered by human activity. Such policy statements precipitated a decision that National Parks would no longer dispose of solid waste within their boundaries. This has resulted in employing municipal or contract services and disposal at regional landfills.

Banff National Park

Banff National Park differs from Jasper, Yoho, and Kootenay in that the town of Banff is autonomous and has a waste management program separate from the Park. Aside from that, Banff and Jasper are very close in geography, weather and summer park visitation which supports a tourism infrastructure. The waste stream should be very similar. In 1990, the town of Banff sponsored a waste management study to determine if implementing recycling was a viable means of supporting a waste management program. The importance of establishing the nature of the waste stream was recognized; however, the national waste composition data was used to assess the amount and composition of the waste produced in the town. This data does not compare with the figures determined in our waste analysis study. The amount of cardboard was assessed to be 15 - 20% of the waste stream from which an estimate was made as to the gross weight of cardboard that could be expected in one year (Oakley, 1990).

Because of the split between the town of Banff and the Park in handling waste, the recycling program is experiencing difficulty in coordinating waste management. Banff now has to look after its own waste and there was some reluctance from the business people to become involved in a complicated and expensive recycling program (pers. comm., Herman, 1992).

In 1993, another study was undertaken to determine how to reduce waste (Siedel, 1993). The results of this study have been adopted by the town. Banff now proposes to undertake the recycling recommendations put forth in the study, starting with cardboard.

Other materials will be brought on stream for recycling when the cardboard and paper fraction are underway. These include construction and demolition waste, organic material (food and yard waste), glass, metal and plastic. There are long range plans to compost the organic material and purchase a glass crusher to create cullet (finely ground glass) to be used in lining ditches. The plastic component is more dependent on an unstable market and may not be recycled until this situation changes (Siedel, 1993).

Yoho National Park

Yoho is a much smaller park than either Banff or Jasper, with fewer visitors. The majority of people visiting this park tend to stay in campgrounds rather than at commercial outlets. There is one small town (Field) with a population of 300. The Superintendent is in favor of recycling and as a result a waste analysis study has been done. The park had access to the summer data produced in Jasper which they corroborated by checking waste bins to determine if the composition was the same. The Yoho garbage is currently being trucked to a transfer station in Lake Louise from whence it is taken to the regional landfill in Calgary.

This changed as of April 1st, 1993 in accord with the agreement achieved through the Canadian Councils of Resource and Environmental Ministers. At that time the trade waste pit was closed down and all trade waste will be trucked to the Golden Regional Landfill, at personal expense. A transfer station was constructed at the old trade waste site for the handling of MSW. Since this facility was in place, all solid waste has then been taken to Golden. There will be a recycling/composting facility developed to handle this waste. What is not compostable or recyclable will be landfilled (pers. comm., Sime, 1993).

Kootenay National Park

Kootenay National Park is similar to Yoho, only it receives fewer visitors and most of them pass through the park without stopping. There is no town in the park and no internal park recycling program, other than a small paper recycling program in the Park office. Both solid and trade waste is taken directly from collection bins to the Windermere dump. There are no immediate recycling plans beyond the present.

Waterton Lakes National Park

This park has a transfer station where all MSW is taken. It is hauled to the regional landfill at Cardston by the Regional Waste Management Authority in Lethbridge. During the summer this amounts to one to two loads a week; in the winter it is down to one load per month. The park also has a trade waste pit which is controlled, and there is a limit on how much construction waste can be dumped. The pit is small and is filling up, even with the restriction on dumping. Waterton has a townsite, but the permanent population is only about one hundred people. There is no bottle depot and all bottles and aluminum cans are collected by local volunteer groups and taken to the depot near Pincher Creek. There is also an office paper recycling program where, again, the paper is taken to Pincher Creek. Waterton undertook a waste minimization study in 1993, the results of which are not yet available.

Waste Management in Related Communities

The Bow Corridor

The Bow Corridor takes in Exshaw and Canmore, plus all development therein, to a boundary a few miles east of Bow Valley Provincial Park. Currently a study is being done by Stanley and Associates to develop a waste management program for this area. Canmore is now forming a Waste Management Committee to investigate its waste stream and recycling options. MSW is presently collected door to door by Bighorn

Transport and taken to the Calgary Regional landfill; trade waste is taken to a trade waste pit in Exshaw. This pit is fenced, has a weigh scale and is manned. There is a user cost to dispose of material here.

There is no bottle depot in Canmore and the locals must take all recyclable material to other centres. Exactly what type of waste minimization program will be developed here, or its connection to Banff, will depend on the recommendations of the Bow Corridor study and the Canmore Waste Management Committee. No programs are in place, either in Canmore or Exshaw, and none will be developed until recommendations are presented through this study.

The Yellowhead Corridor

The principle towns in the ID-14 township, which constitutes the Yellowhead Corridor, are Hinton, Edson, Grand Cache, Robb and Jasper. Outside of the recycling programs undertaken in Jasper, there are no major waste minimization programs in place in these towns. UMA Engineering from Edmonton is currently doing a waste stream analysis of the whole area as well as a survey of what equipment is already in place for future recycling endeavors. The towns also plan to initiate a strong waste reduction, recycling education program for the public. Recommendations and costs will be forthcoming.

SIMILAR COMMUNITIES IN ALBERTA

Riley, Alberta

Riley has long been cited as one of the first communities to experiment with large scale composting. This project was sponsored by Alberta Environment in 1989. The first step was to initiate an education program to ensure a supply of relatively uncontaminated organic waste. This program lasted 18 months. A shredder was built and donated to the project to shred all material into a loose sludge, which was then mixed with a bulking material. The product was windrowed (placed in long columns on the ground) and turned for aeration every three weeks. The project had problems such as plastic bags that were expected to decompose but did not. They also had a disease vector concern and odor problem. The results from this project may be useful to any future composting projects considered for Jasper.

Rimbey, Alberta

Rimby has had an active recycling program operated mainly by the senior citizens, through the Kiwanis Club for several years. They recycle cardboard, paper, glass and aluminum. The town converted an old warehouse into a recycling depot that houses both a cardboard baler and a glass crusher. Because all the work is done on a volunteer basis, they have consistently shown a profit on the cardboard recycling. Rimbey accepts recyclable material from the rural district; therefore handles a considerable amount of material. This operation has been so successful that it is being used as a model for other communities.

Didsbury, Alberta

Didsbury has developed a composting program for yard waste, which may be expanded to all organic waste as their expertise and knowledge develops. The interest in composting comes from the rapid filling up of the landfill the town operates in conjunction with the Regional Waste Management Authority. This landfill will be closed in 1997, and efforts to find a new location have not been successful. The Lions Club is currently undertaking the building of a recycling depot that will handle paper, cardboard, glass, tin and aluminum. The town operates a trade waste pit, but this is not open to the public. All public trade waste is taken to the Regional landfill. The town also recycles used oil, which is taken to Red Deer. Previously, because the oil was mostly contaminated with other oil products, it was accepted by Hub Oil in Calgary at a charge to the town of 35 cents a gallon. Now, however, it is accepted by Turbo Resources Ltd. at no charge.

Brooks, Alberta

Brooks currently has a yard waste composting program. They plan to expand this to include food once the program is established. The initial problem was finding a suitable site that met the criteria set by the Public Health officials. Because of leachate problems, soil profiles were taken at each tendered site to ensure proper containment of the leachate and methane gas. The town has three collection sites strategically located in the town to maximize pick-up. The material is then transported to the composting site where it is processed through a tub grinder and deposited in windrows for turning and decomposition. Other aspects of a recycling program are still being developed.

Other Communities

High River and Okotoks, Alberta, have developed recycling centres. However, a search for a suitable site and the cooperation of the community members has come at a cost. Both these communities experienced difficulty in establishing these centres and in some cases the issues were highly emotional (Seidel, 1992). Alberta Environment cited other communities, such as Dog Pound and Sundry Alberta, that are just beginning to research the establishment of recycling centres and their own waste stream. The town of Stettler had a waste stream analysis done similar as to that in Jasper, but not as extensive. The study was conducted by Dillon and Associates (study not yet released). To Dillion and Associates knowledge, the information has not been put towards a recycling or waste minimization program. Seidel, 1992, concluded that the process of establishing a recycling program can be very complex. The problems encountered by these small towns should be considered to increase the awareness of the process involved and the difficulties that may arise.

4 Jasper Study Methods

The methodology for our study included:

1) a review of the literature dealing with waste stream analysis;
2) consultation with the experts;
3) explanation of the methods used to obtain the data to determine the nature of the waste stream; and
4) analysis of results.

LITERATURE REVIEW

As the cost of handling solid waste increases and available land for landfilling decreases, thus many communities in Alberta are seeking to implement a viable recycling system to reduce their waste (*Recycling of Wastes in Alberta, Technical Report,* 1987). One of the most frequently neglected steps in this development is determining the size and composition of the local waste stream. There is a conviction that data available in the literature dealing with other communities is representative of all communities (Savage et al., 1985). A literature survey supports the need to determine the characteristics of the waste stream for individual communities, for both the practicality of establishing a cost effective recycling program and determining what is going into the landfill (Wright and Neville, 1981).

Van den Broek and Kiror (1969) noted that solid waste is a complex and variable material and cannot be defined readily in one or two simple parameters. There are many variables which affect waste generation and composition. Local waste varies with geography, climate, income level, local by-laws, seasons and economic conditions; to name a few. These conditions change from year to year, making it difficult to predict a given waste stream from a computed model (Klee, 1980).

The need to develop a waste stream analysis for individual communities is important in establishing a proper waste reduction program, but this information can be difficult to obtain (Free, 1986). The most pragmatic approach to a true estimation of the composition and amount of waste produced by any single community is direct measurement at the disposal point (Klee, 1980). The actual separation of waste has not been undertaken by many communities, as it is costly and time consuming. There is also a lack of uniformity in sampling size and categories selected (Howlett, 1990).

Since solid waste is so variable, it is important to select an adequate sample size (Bird and Hale, 1976). The immediate problem facing the researcher in determining the composition of the waste stream is the volume of waste produced compared to what is feasible to sample. The challenge, therefore, lies in devising techniques to overcome the difficulties in separating a small sample from a large heterogeneous mixture (Bird and Hale, 1976).

At the time of Bird and Hale's (1976) study, all forms of garbage were accepted at these dump sites, varying from dust to appliances, with the physical state running from liquids to solids. Since they were working at many different dump sites, there was a need to decide the total size of the sample to be analyzed at any one location. Ideally, the sample should be as large as possible, but practical considerations limit what it is possible to handle. There was a choice of taking a small sample from every truck, as opposed to obtaining a larger sample from a few trucks sampled in a systematic sequence. The latter alternative was preferable to obtain a more comprehensive sample.

For the larger sites, an upper limit of 1,200 lbs per day (544 kg) was set. Samples of general municipal refuse were taken over a 12 month period. The field sample of at least 400 lbs (181 kg) was taken by selecting 25 small samples of 15 to 20 lbs (6 to 9 kg) from each eligible truck. It was the first attempt in Canada to categorize waste on a national scale (Bird and Hale, 1976).

Sample Size:

The intent of this study is to produce reasonable estimates of the composition of waste generated in Jasper National Park based on representative samples. To ensure that these estimates would be as accurate as possible, it was first necessary to determine the proper sample size and number of samples to be taken. Once the data was collected, estimates of waste composition were computed based upon the Student's distribution, which assumes that the population being considered is normally distributed.

The argument for assuming that the percentage of each component being considered follows a normal distribution can be traced through Curruth and Klee (1969), Klee and Curruth (1970) and Klee (1980). In order to evaluate this data, the weights of each component in a sample are divided by the total weights of each sample, resulting in a percentage value for each component. By doing this, the percent of a given component can be compared between different samples (Curruth and Klee, 1969).

In theory, a set of mutually exclusive, collectively exhaustive components, converted to percentages, follows a multinomial distribution (Benjamin and Cornell, 1970; Klee and Curruth, 1970). However, the distributions of individual percentages cannot, in general, be expected to follow a normal distribution, although a normal distribution is considered a good approximation when the percentages lies between 30 and 70 percent. Curruth and Klee (1969) stated that a good normalizing transformation equation for waste component percentages is:

Y = 2 arcsine X (square root of X)

where X is the original percentage and Y is the transformed value.

Klee and Curruth (1970) discussed the problem of using either the multinominal distribution or the arcsine transformation to analyse waste composition data. In both techniques they determined that the number of samples required to accurately determine waste composition was too high to be useful in this type of sampling (over 300 samples per source). The realistic constraints of sampling at this level are well beyond the economic consideration of most organizations, as this is a highly labour intensive exercise.

Klee (1980) concluded that a much more practical, i.e., smaller number of samples can be taken if it can be assumed that each component is normally distributed without prior transformation. He cites a study by Britton (1972), which found that for 90 kg (200 lb.) samples, food waste, paper, metal and glass were all normally distributed. They further tested the normality assumption by conducting random computer sampling of the population distributions of various waste components based on Britton's original data (Klee, 1980).

For elements in the waste stream that appear more infrequently, Britton (1972) took a larger sample size of approximately 140 kg (300 lbs.) A basic distribution is needed for computer sampling; therefore, the studies were based on Britton's original data. Random sampling from these distributions was performed using a digital computer which generated distribution averages of four, ten and twenty samples. For averages of four, the larger components of the waste stream were found to be normally distributed. Britton's samples of wood and textiles were low and a normal distribution was not obtained without taking much larger sample sizes. Klee (1980), however, concluded that for practical purposes a normal distribution would be quite adequate for resource recovery estimations.

Klee and Curruth (1970) determined the sample weight by dividing the samples into three weight categories ranging from 1,400 to 1,700 lbs (635 to 771 kg), 700 to 900 lbs (317 to 408 kg) and 300 lbs (136 kg) and analyzing the variance in each group. Statistical analysis revealed there was no difference in the coefficient of variation between the three groups. Under 200 lbs (90 kg), the variance increased rapidly and over 300 lbs, it decreased very slowly. Therefore, a small sample has the same reliability as a large one in computing back to the population. A 200 lb (90 kg) sample is an optimum weight in terms of cost and efficiency and has been used in most studies. Klee (1980) recommended working with the smaller sample size as source separation could be accomplished with greater accuracy and lead to fewer errors, which magnify when dealing with a large population (Klee, 1980).

The type of sampling found to yield the greatest results was stratified random sampling. Once generation sources are established, the material may be broken down into the categories most representative of the community. Klee (1980) recommended sampling over a period of one week to incorporate the daily fluctuation expected over this time.

CONSULTATION WITH THE EXPERTS

To successfully study Jasper's waste stream and transportation costs, many people involved in these two fields were consulted. No detailed information was available on the waste stream of the Jasper community prior to this study. Thus it was necessary to obtain much of the material from persons working in this field prior to this time. Experts consulted included government officials, university professors, interest groups and various consulting companies, all of whom kindly donated their time and resources to this study. In the field of transportation, much information came from private companies both in the trucking industry and in the recycling industry. Stanley and Associates were particularly helpful in establishing the procedure for sampling a large amorphous population of garbage such that the samples would reflect the true nature of composition and quantity.

METHODOLOGY USED IN THE JASPER STUDY

The sampling technique in Jasper was based on that used in Ottawa by Stanley and Assoc. (1991), as established by Robertson and Bertrand (1991). The Ottawa study sampled over a period of one week for each season. Their total sample size did not exceed 25 truck loads during each sample session. The samples were selected randomly from the contents of a full truck by the use of a front end loader. Each sample was kept within the 200–300 lbs. (90–136 kg) range recommended in Klee's (1980) study. The material was dumped from the truck onto the floor of the transfer station and then picked up with the loader and deposited onto a tarp.

Because of the size of the project and the manpower available, the material was sorted down to super fines (the small particles left after sorting the major components). Stanley and Associates sampled over a period of one week during the spring, summer, winter and fall seasons. The purpose of the seasonal sampling was to detect the variation expected from seasonally dependent elements over the course of a year (pers. comm., Metcalf, 1992).

In Jasper, the analysis of the waste stream lasted over a one year period, in that samples were taken for 12 months. The study, however, was spread over a two year period. The first samples were taken during the summer of 1991 (June, July and August). The winter samples (December, January, and February) were taken in 1992; however, January and February samples were taken at the beginning of 1992, where as December samples were taken at the end of 1992. Spring (March, April, and May) and Fall (September October and November) samples were taken in 1992. The primary study was done at the transfer station, where solid waste was weighed and sorted into 13 categories.

The second part of the research consisted of determining the quantity and type of material going into the trade waste pit. Early studies at the trade waste pit revealed that comprehensive data was not available on the amount of material being dumped. There where no scales to weigh the material nor was there any control of what was dumped or when.

Table 3 Materials Identification – Transfer Station

List of Source Materials	
Organics:	Food Wastes All material classified as staple foods.
Organics:	Yard Waste All organic material classified as plant material.
Organics:	Wood All wood products.
Organics:	Other This includes textiles, leather, rubber, diapers and other sundry organics.
Paper:	All paper products. These include fast food paper products.
Cardboard:	All cardboard products, including packaging material.
Plastics:	All plastic material including styrofoam, and non-paper packaging material.
Glass:	All glass products.
Metal:	All metal products. These include objects primarily made of metal, such as irons, toasters, etc.
Aluminum:	All aluminum products (cans).
Mineral fill:	Ash, dirt, gravel, etc.
Mineral Other:	Construction debris, ashphalt, dry wall, ceramics
Hazardous Waste:	Propane, gas, household hazardous waste, batteries.

Project Description

The categories selected for separation were determined from those used in the Ottawa and Edmonton waste characteristic studies (Table 3). Once the basic categories were established, they were reviewed with park personnel who had been involved in the field. The categories remained essentially the same as those established in other studies, but, took into consideration such materials as 'ash' which had been observed to be high in the past. Cardboard was classed separately from paper as it is a major component of the waste stream and the market for recycling is separate from that of paper.

The transfer station in Jasper handles all the solid waste in the park. It is brought in by the park garbage trucks, dumped onto the floor, and pushed into a transfer trailer with a front end loader. The refuse is then hauled to the Hinton Landfill. Each truck that comes in has a number and a set weight when empty. The source of garbage each truck hauled was classified as either commercial, residential, campground or mixed. Mixed garbage consisted of an east or south highway run that picked up both commercial and campground waste and was therefore weighed but not sorted. As specified by Bird and Hale, it is not possible to obtain a clear separation nor an accurate determination of the mix (Bird and Hale, 1976). During the sampling period all the trucks were weighed on commercial scales located at the entrance to the sanitary landfill.

Three sources (generators) of waste were selected to be sorted daily. The average amount of garbage sorted was approximately 700 lbs. (317 kg) per day. Since it was expected that there would be differences throughout the week, an eight day sampling period was selected for June, July and August. This was a total of 24 truck loads, from each source, over the course of the summer. This exceeds the required 16 samples per source when considered over three months, but is less in the duration of one sample period. Essentially the sample size per season corresponded with the Stanley study, (1991) but was spread out over a 3 month period. It should be noted that in Klee's (1980) study much of his work was done at municipal dump sites, which were open to the public; therefore, the truck size became an important factor as many of the trucks selected were light. A smaller sample had to be taken from trucks carrying less refuse than the large commercial trucks. At the Jasper transfer station all trucks coming in throughout the summer were the same size and weight, and only trucks that were fully loaded were sampled.

The selected trucks dumped their load onto the transfer station floor, which was then mixed with a front end loader, from which a randomly picked load was deposited in a wire mesh bin. The bin was used, rather than a tarp, as wind blowing through the station would otherwise scatter the refuse and confuse the sorting. The bin was built with two bays allowing for two sorters to work at once.

It was difficult to estimate the weight of each front end loader scoop as density and weight varied with the weather conditions, and the source of the garbage. Thus, to achieve the two hundred to three hundred lbs, each sorter would start sorting strictly from the top down. If a bag was selected it was sorted completely into the categories specified. Much of the garbage was loose, either from originally being disposed in this manner or from split bags. The bin was useful in containing this loose material, and it was sorted as it was encountered.

Once an item was classified such as ceramics, it was consistently sorted into that category. If an item was a mixture of two components i.e; plastic and metal, it was broken down into the two components. In some instances a complete breakdown was not possible. This occurred rarely, and the object was classified in the category most representational of its make up. The greatest problem that affected the results was inadvertently coming across a large quantity of material that belonged to a category that usually yielded low figures. For example, a large quantity of metal might be found, when normally very little was present. It was important to have an experienced person present during the sorting to oversee the work of the employee. Each basket was checked for misplaced items before being weighed.

Again each truck was assessed as to its source, and only those containing a pure source of garbage was sampled. Residential and commercial loads were randomly selected throughout the week as there were enough incoming trucks to allow for a change in schedule. Thus, one morning residential would be selected and commercial selected mid-day or early afternoon. As there was only one pure campground load each day it was sorted when it arrived. The composition of campground garbage was consistent throughout the week, but residential and commercial garbage composition varied considerably, depending on the time and the day a truck was selected. It was important to alternate truck selection to avoid bias. Should a run from all the bungalows be exclusively selected, results would be skewed in the direction of food. If, however, only IGA and Super A routes were selected, the results would swing to a heavy cardboard element. Similarly, residential weights could be skewed toward yard waste by sampling truck routes that serviced only single family dwellings. The trucks selected alternated at random with those trucks picking up refuse from multi-dwelling establishments (apartments) where smaller amounts of yard waste are produced. If the summer were to be considered one season, each source would have a total sample size of 16. Since this exceeds the sample size calculated by Klee (1980), individual categories were not selected for calculation of sample size.

The raw data consisted of recording a series of individual weights taken from one sample placed in the wire mesh cage. The weights were obtained by separating the garbage (starting randomly from the top) into one of the thirteen designated categories. The selected material was placed in a container labelled for that category. Once the container was full, the contents were weighed and emptied. Not all the containers would be full, as some categories such as metal would not amount to that much. This process was repeated until the required sample size was achieved, usually in excess of 200 lbs. A large component of the waste stream would be weighed more often than a small component (such as metal) because it would be encountered more frequently. Each of the individual weights were then tallied to give an overall weight for that day. The daily tallies were entered on a master sheet at the end of the eight day sample period, and a grand total was calculated for each of the thirteen categories. These totals were then converted into a percentage of the total amount of garbage against the total amount of garbage shipped to Hinton throughout the sample period. (The degree of accuracy is given for each season at the end of the review of waste composition, Section 5.)

Comparison between each source, category, and month could then be made. As noted, some of the variations were due to a significant increase or decrease in one category. Weather also played a role in altering the weight expectations in certain categories such as cardboard, which absorb moisture. The calculations and significance of the sampling remained unknown until the end of the sample period so as not to bias these results by trying to achieve an expected weight in a certain category.

Generation Rates

To arrive at a valid extrapolation of the quantity of waste generated in each category it is necessary to record the weights of the trucks and contents over the sample period (Savage, et al., 1985). In this study, all trucks were weighed during each sample period and the weights recorded with respect to the source from which the waste was generated. During the course of this study the trucks were weighed on two different sets of scales. The summer weights were taken on a set of scales located 200 yards (182 m) from the transfer station. These scales were drive on and the weights were recorded in pounds. During the summer there were eight trucks in use, each making several trips a day. This high activity made it difficult for individual sorters to record each weight considering the distance from the transfer station to the scales. These weights were recorded by each driver as they came in with a load.

The other set of scales was located within the transfer station itself. These scales weighed the amount of garbage that left the transfer station in the transfer trailer daily. During the spring, fall, and winter seasons, the amount of garbage was reduced and these scales were free to weigh the incoming trucks. These weights were recorded in kilograms.

5 Review of Collected Yearly Data

The results of the waste sampling are presented for each season, by source, and for each category. The generation rate or quantity of total garbage produced is also discussed for each season to allow for a summation of the estimated amount of material expected from each category and from each source. The four seasons of the year were allocated as follows:

Summer: June, July, August
Fall: September, October, November
Winter: December, January, February
Spring: March, April, May

DATA COLLECTED SUMMER, 1991

Commercial Waste Summer 1991

Most of the commercial garbage in Jasper is generated from bungalows, hotels, restaurants and retail stores. The trucks which pick up this garbage also pick up CN waste; however, this waste tends to be more commercial than industrial. There is an industrial park in Jasper, but few of the businesses produce industrial waste. Construction waste is not normally dumped here as the design of the holding bins does not permit deposition of large solid objects. Material of this sort is hauled to the trade waste pit.

Food:

Commercial garbage differs from residential and campground garbage in that food constitutes up to 37% of the waste stream (Figure 3). This compares with an average of 14% in residential waste and 13% in campground waste (Figure 5). There seems to be a lot of discarded food from restaurants and retail outlets such as IGA and Super A. Food waste is

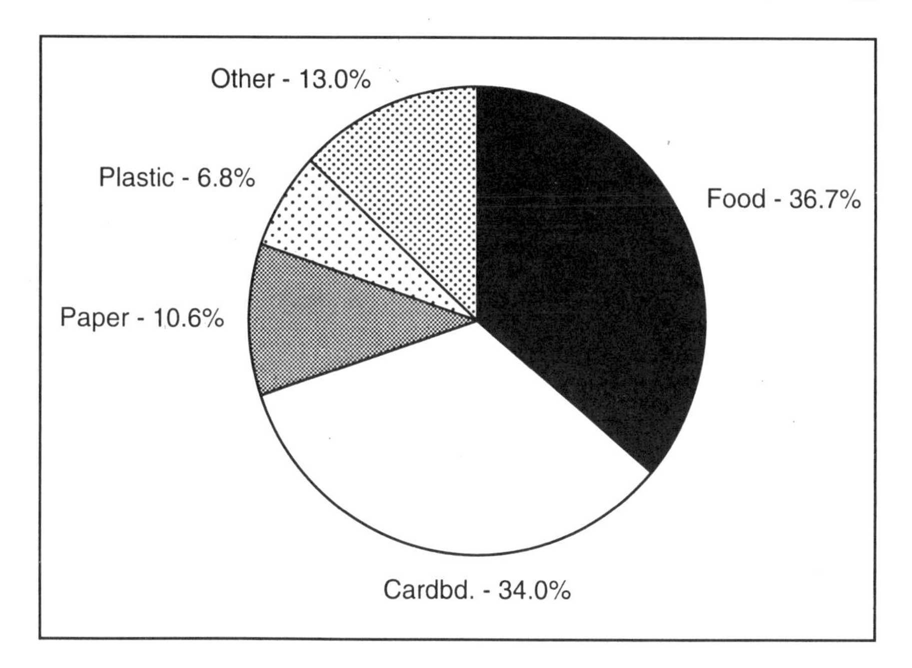

Figure 3 Commercial Composition Summer – 1991

particularly high from the Jasper Park Lodge. The food waste in JPL's garbage could be as high as 50% of their garbage.

Yard Waste:

Commercial refuse is lower in yard waste than residential waste, as these outlets do not have yards. Some of the motels and bungalows do have some landscaping, but little of this showed up at the transfer station. There is quite a difference in proportion: 1.4% commercial yard waste compared to 26.5% residential yard waste.

Organics Other:

These items are defined in Table 3 and consist of organic material other than food or yard waste. This is not a large component, by percentage, in commercial waste. Aside from the occasional organic fibre, rugs, etc., thrown out by JPL, little organic waste was found.

Paper:

Paper was low in commercial garbage, but, as most of it consisted of accounts and files, it may be that much of the desk work is put on hold untill the tourist season is over.

Cardboard:

Cardboard is the most visible waste in the waste stream. Before any weighing occurred, observation led many of the collectors to think it was the dominant material in the waste stream. Although cardboard undoubtedly comprises the greatest amount by volume, it is still lower by weight than food (Table 4).

Plastic:

Plastic is a very light material and for this reason does not figure highly in the waste stream. In recent years manufacturers have been making plastic bags and containers thinner; therefore, the amount of plastic by weight has not increased as much as expected (Rathye, 1991). Also, there is a bottle depot in Jasper that recycles plastic bottles, which eliminates some of this material from the waste stream.

Table 4 Monthly Composition Analysis for Summer 1991

	Commercial					
Material	**June/91**		**July/91**		**August/91**	
	lbs.	%	lbs.	%	lbs.	%
Food	735.6	39.8	873.4	44.0	490.2	26.3
Yard	5.4	0.3	0.3	0.0	74.3	4.0
Wood	20.4	1.1	8.2	0.4	24.8	1.3
Org. Oth.	34.4	1.9	19.5	1.0	27.4	1.5
Paper	223.3	12.1	199.8	10.0	164.3	8.8
Cardboard	493.8	26.7	601.6	30.2	798.1	42.9
Plastic	144.4	7.8	114.7	5.8	120.5	6.5
Glass	78.2	4.2	62.6	3.1	87.1	4.7
Metal	64.9	3.5	35.5	1.8	39.8	2.1
Aluminum	23.9	1.3	15.0	0.8	14.8	0.8
Min. Mix	15.7	0.8	2.0	0.1	1.7	1.0
Mix other	0.0	0.0	56.1	2.8	17.6	9.0
Hazardous	7.9	0.4	0.4	0.0	1.0	0.1
lbs.	1847.5	99.9	1989.1	100.0	1861.4	100.0
Kg	839.8		904.1		846.1	

Glass:

Glass only showed up in significant amount, from lodges after a long weekend. Glass from this source is being recycled through the bottle depot, which results in a considerable diversion of this material from the waste stream. Glass was highest in campground garbage, where the opportunity for recycling is the lowest. There is a recycling centre in Whistlers Campground, but it is some distance from many campsites and difficult to locate. For effective recycling, more depots would be required.

Aluminum, Mineral Mix, Mineral Other and Hazardous Waste:

These four categories make up the lowest group in the commercial waste stream. Aluminum is very light, and much of it is diverted from the waste stream through recycling. mineral mix, mineral other and hazardous waste are also low in the waste stream as commercial garbage is dominated by food and cardboard waste (Figure 3). Most construction waste is taken to the landfill, as is gravel and dirt. This material does not often show up at the transfer station. This is also true of hazardous waste.

Residential Waste Summer – 1991

Residential garbage (Table 5) was collected from both the east and west ends of town. The trucks had to be carefully monitored to determine the pick up route, so as to not select the run which included the schools and the town's activity centre. These runs also were divided into a normal residential district comprised mostly of individual homes, and multi-dwelling accommodation, including apartment buildings.

Food:

The food component was not overly high in proportion to the rest of the waste stream. People at home throw away less food than commercial outlets.

Yard:

Residential garbage was more evenly distributed with the exception of yard waste, which tended to be higher in June. Yards are still being groomed in June, which accounts for the higher values in this category. This may vary from year to year, depending on the weather.

Wood:

This catagory is consistently low throughout the summer.

Table 5 Monthly Composition Analysis for Summer 1991

Residential						
Material	**June**		**July**		**August**	
	lbs.	%	lbs.	%	lbs.	%
Food	285.2	15.7	251.7	14.1	226.6	12.2
Yard	616.5	33.9	370.6	20.7	461.4	24.9
Wood	28.3	1.6	72.3	4.0	30.3	1.6
Organics Oth	149.1	8.2	153.4	8.6	148.9	8.0
Paper	269.0	14.8	270.0	15.1	308.4	16.7
Cardboard	183.1	10.1	269.4	15.1	321.5	17.4
Plastics	104.0	5.7	124.4	7.0	100.3	5.4
Glass	79.6	4.4	84.3	4.7	115.6	6.2
Metal	56.9	3.1	103.3	5.8	73.8	4.0
Aluminum	18.2	1.0	26.8	1.5	30.8	1.7
Min. mix	11.0	0.6	50.4	2.8	21.4	1.2
Min. other	16.2	0.9	9.6	0.5	11.7	0.6
Hazardous	2.5	0.1	2.3	0.1	1.4	0.1
lbs.	1819.6	100.1	1788.5	100.0	1852.1	100.0
Kg	827.1		813.0		841.9	

Organics Other:

Organics Other in residential garbage consists mainly of disposable diapers and discarded clothing. Other contributors would be discarded rugs and curtains, rubber mats and other household organics, other than food. This material is a considerably higher component of the waste stream in residential waste than in commercial waste (comparison of Table 4 with Table 5). It did maintain considerable uniformity throughout the three summer months.

Paper:

The Paper component is a conglomerate of every type of paper discarded but, reflects to a considerable extent, the amount of packaging waste found in residential garbage. Paper packaging is not found to this extent in commercial garbage as most of the food comes in bulk and in large cardboard boxes.

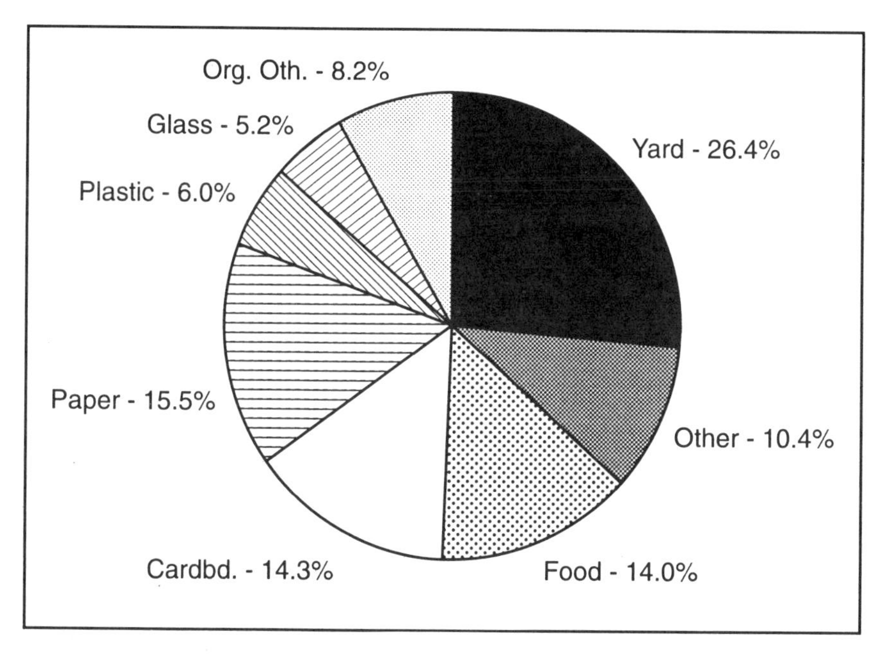

Figure 4 Residential Composition Summer – 1991

Cardboard:

Cardboard, similarly consists mostly of packaging found mainly as small light boxes. Large heavy packing boxes are not found to the same extent as is found in commercial garbage. It takes a considerable amount of the light cardboard to add up to a significant weight. The weight in cardboard also increased in August as it did in both campground and commercial garbage. The apparent increase in the weight of the cardboard reflects the increased absorbtion of water in the cardboard in August.

Plastic, Glass, and Aluminum:

Plastic and aluminum, due to their being light material are not a large component by weight of Residential waste. This holds true for Commercial waste also.

Glass is recycled to a great extent. It is usually stored in the basement and then picked up by bottle drives or driven off in bulk to a bottle depot. Aluminum may be recycled, but, the amount found in the waste stream does not vary much from either campground or commercial garbage, perhaps because it is simply a light material.

Mineral Mix, Mineral Other, and Hazardous Waste:

Again these components are low in the waste stream and show considerable uniformity from month to month, with the exception of Mineral Mix, due to the deposition of a large bag of kitty litter and light gravel in July.

Campground Waste Summer – 1991/92

The campgrounds in Jasper are only open for four months of the year, during which there is a sufficient amount of garbage generated to reflect the true amount of waste contributed from this source. For this reason the campground data has been lumped into one general summer season which includes June, July, August and September. June was sampled in both 1991, and 1992, because the first sample taken in 1991 was during a relatively early and unusually wet season. The seasonal sample taken in June 1992 is believed to be more representive of the type of waste generated at a campground at this time of year. Results indicated in Table 6, Figure 5, that campground garbage is similar to residential garbage (Table 5) in distribution between the categories, with the exception of minerals other, which is composed mainly of campground ashes.

Food:

The percentage of food waste in June, of 1991, was very low and may have been affected by the amount of collected yard waste which ceased in July and August. (The campgrounds are still being groomed in June for the summer but grooming stops once the campgrounds are prepared). Examination of the original June data showed consistently low amounts of food through all eight days of sampling, indicating that in the random samples the food never appeared to any great extent. The campground was not very full at this time and the weather was cool. This may have promoted the eating of fast foods, from which there was little food waste as opposed to cooking regular meals. A second campground sample was taken in June, 1992, again, over an eight day period. However, it was taken at the end of the month in good weather when the campground was full. The amount of food discarded came closely in line with that of July, August and September. The average of the two months is given in Table 6.

Table 6 Monthly Composition Analysis for Summer 1991

	Campground							
Material	June		July		August		Sept.	
	lbs.	%	lbs.	%	lbs.	%	lbs.	%
Food	260.9	15.5	362.8	19.4	250.8	14.3	166.6	15.5
Yard	84.1	5.0	0.1	0.0				
Wood	13.2	0.8	2.0	0.1				
Org Oth	36.8	2.2	77.5	4.1	87.0	5.0	14.5	1.4
Paper	204.2	12.2	283.5	15.2	288.2	16.5	147.3	13.7
Cardbd	102.7	6.1	159.6	8.5	198.0	11.3	60.7	5.7
Plastic	125.0	7.4	179.5	9.6	105.4	6.0	94.0	8.8
Glass	156.2	9.3	276.3	14.8	183.5	10.5	134.6	12.5
Metal	54.1	3.2	90.6	4.8	85.6	4.9	64.5	6.0
Alum.	31.0	1.8	74.5	4.0	35.3	2.0	25.5	2.4
Min. Mix	613.1	36.3	336.2	18.0	500.5	28.6	363.3	33.8
Min. Oth	2.3	0.1	20.7	1.1	12.8	0.7	2.5	0.2
Haz.	3.3	0.2	7.7	0.4	4.1	0.2	2.0	0.0
lbs.	1686.9	100.1	1871.0	100.1	1751.2	100.0	1075.5	100.0
Kg	748.0		850.5		790.0		487.8	

Yard Waste:

In June yard waste similarly varied from the 1991 sampling session to the 1992 session. Yard waste was being cleared up at the beginning of the month in 1991, but was completed in 1992 by the end of the month. The sampling results in June 1992 agreed with the results from July and August, indicating that the early sample in June, 1991, was not indicative of true campground garbage once the season began. The September sampling results were very much in line with results compiled for July and August.

Organics Other:

The organics other category was lower proportionally in the campground waste stream, but was still high enough to suggest that the majority of visitors to the campground reflects a strong family element. Diapers were common, but such things as rugs, boots, rubber mats, awnings and clothing were also found. The average amounts remained the same throughout the summer, while the campgrounds were fully operational.

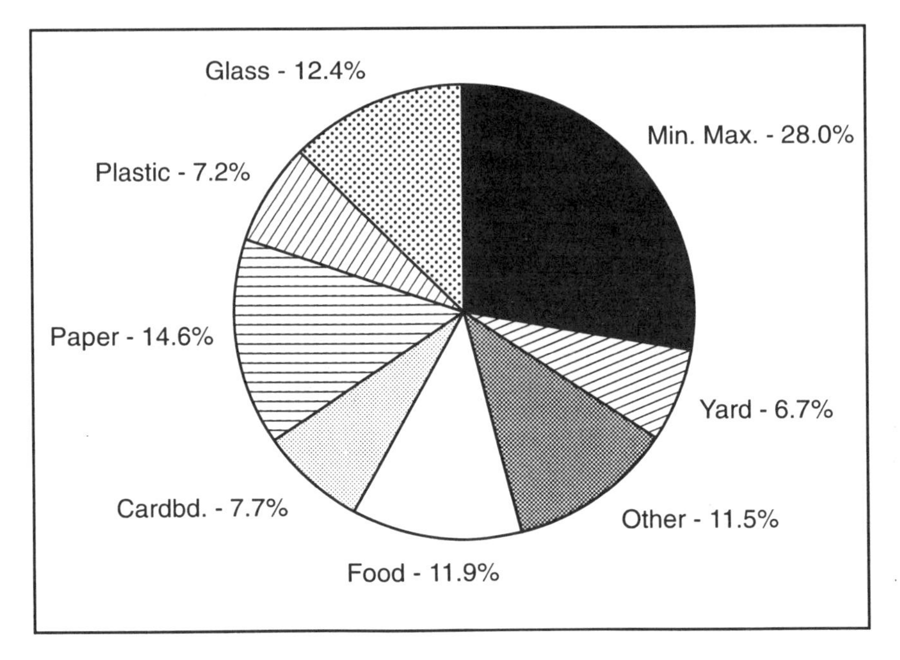

Figure 5 Campground Composition Summer – 1991

Paper:

Paper was similar to that of residential waste and remained steady at an average of 15% throughout the summer.

Cardboard:

There was an increase in cardboard in August, similar to the increase in commercial garbage. Again, this was probably affected by the heavy rainfall experienced during this sample period. Cardboard was a smaller proportion in the campground waste stream than either commercial or residential garbage.

Glass:

The percentage of glass was quite high compared to commercial and residential garbage, suggesting, more food and drink in glass containers were purchased and not recycled. Lack of readily available water in a tap (which requires hauling) may explain this. There is a recycling depot in the campground, but it is inadequate to the needs of the campground and is quite far from most of the camp sites.

Table 7 Average Percent Composition by Weight for Each Source Over Summer Sample Period

Material	Commercial	Residential	Campground
Food	36.8%	14.0%	13.9%
Yard	1.4%	26.5%	4.4%
Wood	0.9%	2.4%	0.2%
Organic Oth.	1.4%	8.3%	3.4%
Paper	10.3%	15.5%	14.2%
Cardboard	33.2%	14.2%	7.4%
Plastic	6.7%	6.0%	7.5%
Glass	4.0%	5.1%	11.6%
Metal	2.5%	4.3%	4.2%
Aluminum	0.9%	1.4%	2.4%
Min. mix	0.3%	1.5%	30.0%
Mix. Oth	1.3%	0.7%	0.5%
Haz.	0.2%	0.1%	0.2%

Plastic and Aluminum:

Aluminum and plastic percentages closely resembled those of residential and commercial garbage. Aluminum was a lower commodity in the waste stream. This may be due to the recycling depot in the campground, but aluminum is also on average a lighter item than plastic, although both are light by weight.

Mineral Mix:

Mineral mix (ashes) was highest in June, dropped by 50% in July and rose again in August. Because the campground was not full in June all the ashes in the campground were placed in the green bins. During July and August a special truck hauled one load of ashes a day and the surplus was placed in the bins. The increase in August probably reflects the increased visitation during this month creating a greater amount of ashes. The trucks carrying ash alone were marked separately from the regular campground hauls. September was similar to June, probably due to continued high use and lower temperatures.

Mineral Other:

This is largely construction material of which there was very little in campground waste.

Hazardous Waste:

Hazardous waste was more common in campground refuse than the other two sources. The main waste product being discarded was batteries of various sizes, which are heavy items.

Table 8 Truck Weights Summer Sample Period

	June	July	August
Scales	89,945 kg	119,800 kg	142,014 kg
	94.9%	95.2%	86.0%
Trailer	94,780 kg	125,840 kg	165,070 kg
	Scales Total:	**351,759 kg**	
	Trailer Total:	**386,090 kg**	
Percent Accuracy:	**91.1%**		

Generation Rate Summer – 1991

The amount of garbage by weight did not vary much between July and August as these two months are the height of the tourist season (Table 8). Despite using older scales, the percent accuracy was high; the lower weights from the trucks were due to missed weigh-ins by a truck driver at the beginning of each sample session. The major concern is the distribution of garbage between the four sources (Figure 6). During this summer the commercial garbage amounted to 48% of the total park garbage; however, the mixed garbage from outlying areas was also high.

DATA COLLECTED FALL – 1992

The fall data was collected in 1992, when the economy was depressed; however, the weather was similar to that of 1991. The campground data

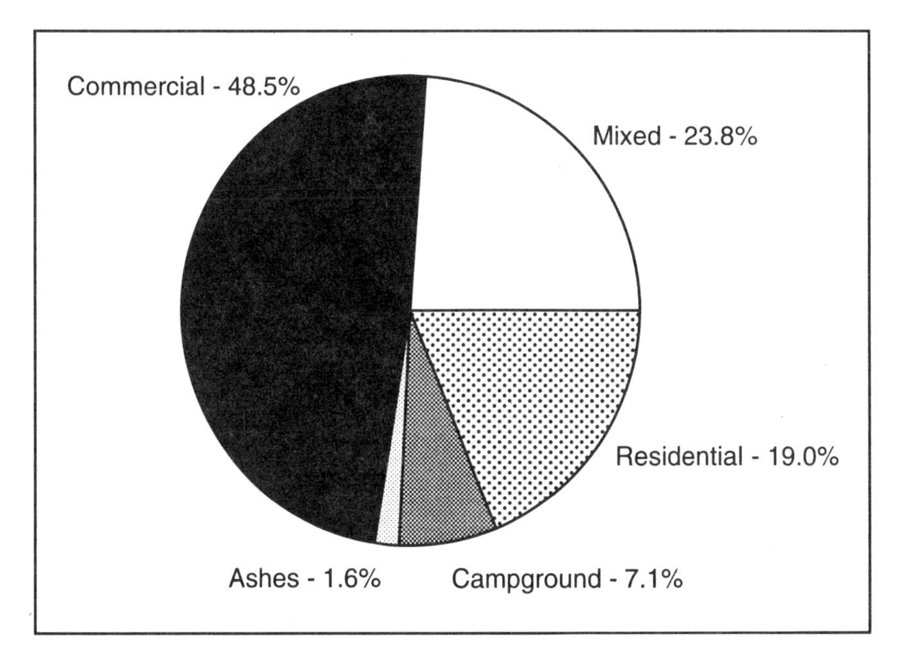

Figure 6 Material Generation Rate Summer – 1991

was only sampled in September, as all campgrounds were closed after this. This data was included in the summer campground calculations.

Commercial Waste Fall – 1992

Food:

Food waste in September and October was slightly higher than the average food waste in the summer. This started to fall in November when restaurants closed during the off season. At this time there is a significant drop in the tourist population and all commercial outlets reduced the amount of food ordered and supplied.

Yard:

Yard waste was never very high in commercial garbage, and was quite low throughout the fall. The highest rate was achieved in September but was still low compared to the rest of the waste stream.

Organics Other:

This organic material, described in Table 3, rose considerably from an average percentage of 1.4% in the summer to a high of 8.4% in November. This compares to a lower rate in the winter of 3.4%. In reviewing

Table 9 Monthly Composition Analysis for Fall 1992

Commercial						
Material	September		October		November	
	lbs.	%	lbs.	%	lbs.	%
Food	735.4	41.5	725.6	40.7	559.7	31.1
Yard	18.9	1.1	4.5	0.3	3.1	0.2
Wood	7.7	0.4	9.0	0.5	14.7	0.8
Org. Oth.	71.5	4.0	29.2	1.6	151.3	8.4
Paper	321.7	18.1	290.9	16.3	321.1	17.8
Cardbd	357.0	20.1	507.2	28.5	504.4	28.0
Plastics	119.1	6.7	108.3	6.1	125.9	7.0
Glass	75.4	4.3	42.7	2.4	52.7	2.9
Metal	40.2	2.3	37.8	2.1	37.4	2.1
Aluminum	21.8	1.2	16.5	0.9	14.6	0.8
Min. Mix						
Min. Oth.	0.8	0.0	9.3	0.5	15.4	0.9
Haz.	4.0	0.2				
Total	1773.5	99.9	1781.0	99.9	1800.3	100.0
Kg	804.5		807.9		816.6	

the daily samples, there seems to be some evidence of renovation, as much of this material was discarded rugs and hotel sheets. Discardment of this material is highest at this time of year.

Paper:

Paper disposal also rises in the fall from an average of 10% in the summer to 17% in the fall. This is the time of year that businesses catch up on the paperwork neglected during the summer when business is at its peak.

Cardboard:

The amount of cardboard drops from 34% in the summer to 24% in the fall and winter. Again, it must be remembered that the summer data was taken in 1991 under wetter conditions than that in the fall of 1992. Also, the tourist trade drops off rapidly in September. The fall data for cardboard comes closer to the winter and spring data.

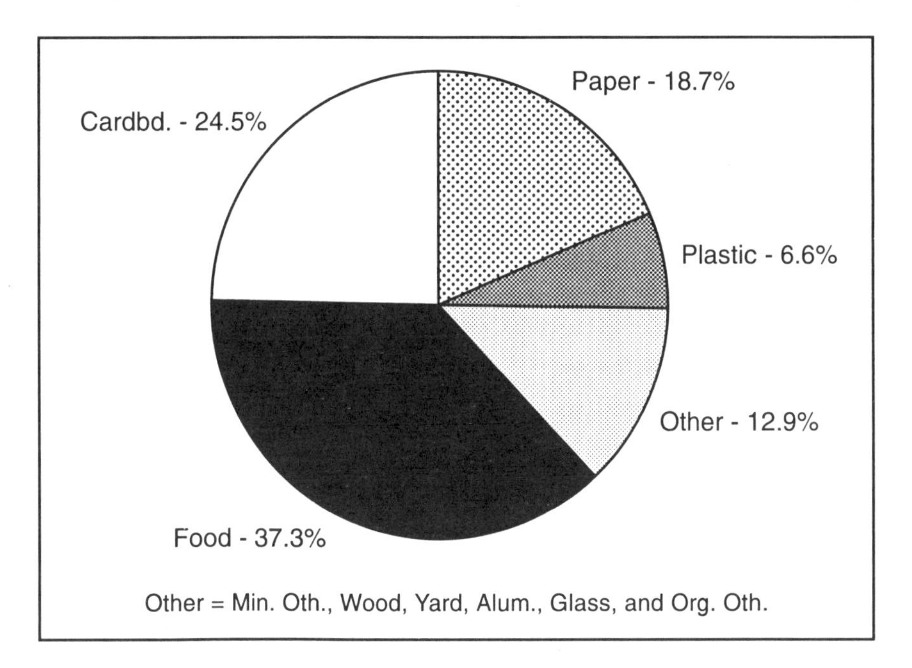

Figure 7 Commercial Composition Fall – 1992

Plastic:

Plastic remained relatively uniform throughout the fall sampling and did not change much from the data taken during the rest of the year.

Glass:

The amount of glass drops from an average of 4% in the summer to 3% in the fall. Winter and spring values are lower yet, with an average of 2%. This may be indicative of the greater reliance on the bottle depot and the fact that the campgrounds, which are a major source of glass, were closed.

Metal and Aluminum:

There was no real change in either of these materials in the waste stream. Both values remained steady on average throughout the sampling period and the year.

Mineral Mix, Mineral Other, and Hazardous Waste:

These materials are again typically low throughout the sample period and are in line with the rest of the year. Mineral mix is not present.

Table 10 Monthly Composition Analysis for Fall 1992

Residential						
Material	September		October		November	
	lbs.	%	lbs.	%	lbs.	%
Food	240.9	15.8	238.2	13.7	185.7	13.8
Yard	327.2	21.4	488.9	28.1	44.9	3.3
Wood	17.6	1.2	16.1	0.9	56.6	4.2
Org. Oth	163.1	10.7	185.2	10.6	199.1	14.8
Paper	232.4	15.2	279.4	16.0	267.4	19.8
Cardboard	242.9	15.9	255.6	14.7	244.6	18.1
Plastic	115.1	7.5	118.6	6.8	120.1	8.9
Glass	67.8	4.4	68.3	3.9	71.9	5.3
Metal	73.8	4.8	65.9	3.8	66.7	4.9
Alum.	18.7	1.2	20.5	1.2	19.4	1.4
Min. mix	24.3	1.6			59.0	4.4
Min. other	2.3	0.2	4.5	0.3	14.3	1.1
Haz. waste	0.6	0.0				
Total						
lbs	1526.7	99.9	1741.2	100.0	1349.7	100.0
Kg	694.0		791.0		613.5	

Residential Waste Fall – 1992

Food:

The amount of food waste in the fall remains consistent with the summer waste. It only changes when the yard waste is no longer present. Proportionately, yard waste is still dominant in the fall.

Yard:

Yard waste drops from that produced in the summer, mainly because of the significant drop in this material in November. Also the yard waste in the fall is dryer and lighter at this time of year. Yard waste achieves its highest value during June and October which corresponds with spring and fall clean-up.

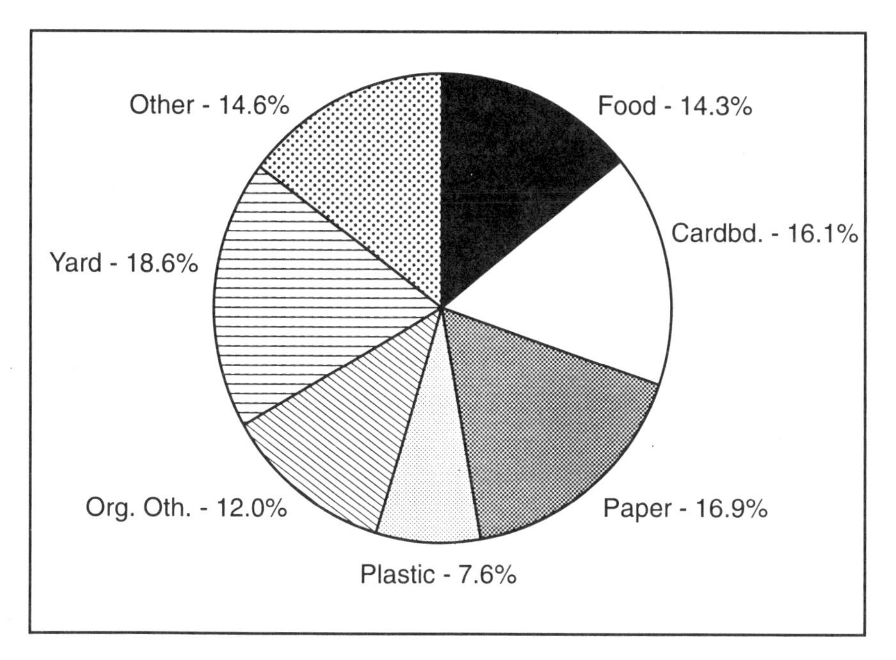

Figure 8 Residential Composition Fall – 1992

Wood:

Wood is a small portion of the waste stream, mainly because most of this waste goes to the landfill. There was an increase in November, however, due to some construction material that was placed in one of the residential bins on two consecutive days.

Organics Other:

This material remained very uniform throughout the entire year. Again, this consists of diapers, discarded clothing and houshold organics such as leather, textiles etc.

Paper:

Paper corresponds with summer volumes through September and October, but rises in November to almost 20% as compared to 10% in the summer. This apparent increase is explained by the dramatic decrease in yard waste and comes closer to reflecting winter data, particularly December.

Cardboard:

Cardboard remained uniform through September and October and consistent with the summer data . In November the percent rises, reflecting the decrease in yard waste. These are consistent with December figures.

Plastic:

Plastic, like many other materials in residential waste does not change significantly from that produced during the rest of the year. There are no major changes in this category.

Glass:

Glass corresponds with the summer data throughout September and October, but, like cardboard and paper, increases in November and is comparable to winter months. There seems to be less inclination to recycle in the colder months, which is probably the reason for the increase at this time.

Metal, Aluminum, Mineral Mix, Mineral Other and Hazardous Waste:

These categories are all very small amounts in the waste stream and none of these values vary significantly from month to month or throughout the year.

Generation Rate Fall – 1992

The fall generation figures were obtained as they were in the winter and spring by making use of the transfer trailer scale. Although the amount

Table 11 Truck Weights Fall Sample Period

	Sept	Oct	Nov
Scales	83,046 kgs	70,120 kgs	59,411 kgs
	99.0%	94.0%	99.0%
Trailer	82,300 kgs	66,900 kgs	59,950 kgs
	Scales Total:	**212,577 kg**	
	Trailer Total:	**209,150 kg**	
Percent Accuracy: 98.0%			

Table 12 Average Percent Composition By Weight For Each Source Over the Sample Period

Material	Commercial	Residential
Food	37.7	14.4
Yard	0.5	18.6
Wood	0.6	2.0
Org. Oth.	4.7	11.9
Paper	17.4	16.9
Cardbd.	25.6	16.1
Plastic	6.6	7.7
Glass	3.2	4.5
Metal	2.2	4.5
Alum.	0.8	1.3
Min. Mix	0.0	1.2
Mix Oth.	0.5	0.5
Haz.	0.1	0.0

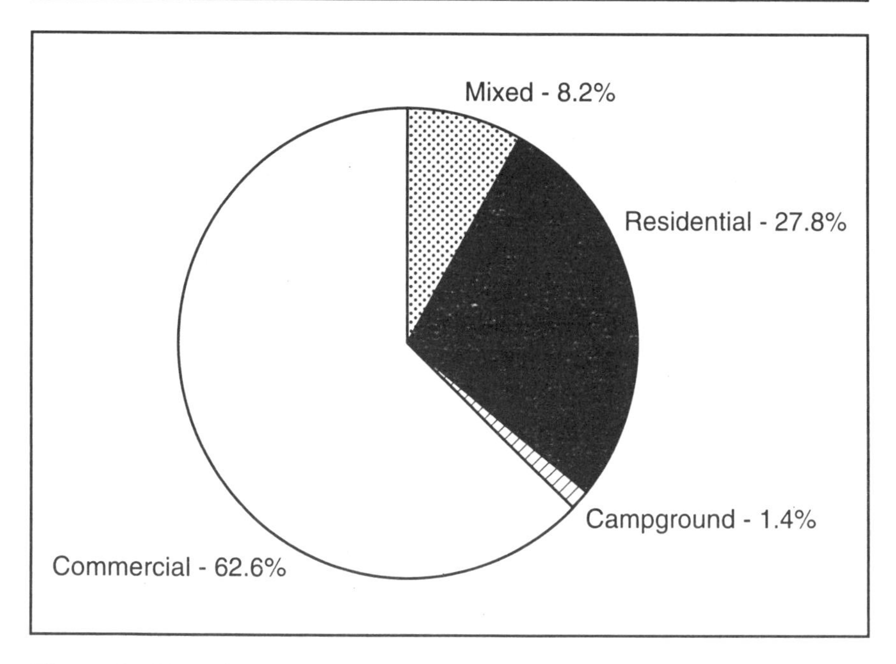

Figure 9 Material Generation Rate Fall – 1992

was high in September, this figure decreased in October and November, but,not enough to lead to an overall decrease between 1991 and 1992. The park was still receiving campground and outlying mixed loads of garbage in September, but this source of garbage was finished in October and November. The ratio of weighed individual loads to that recorded as being removed from the transfer station via the transfer trailer, can be found in Table 11.

DATA COLLECTED WINTER – 1992

The most significant difference from summer to winter is the reduction in the quantity of garbage and the disappearance of campground and mixed garbage runs. At this time, the waste stream is generated through commercial and residential garbage alone. A minor source of waste is contributed by the Marmot Basin ski area, but this was not significant in overall weight.

Because the volume of garbage being trucked to Hinton was reduced, it was possible to pull the transfer trailer off the station scales and weigh each truck as it came in. This was of particular value as the old scales were removed in the fall, and it was possible to record all weights in one building. The truck weights retained the same accuracy (90%) of that in the summer months, but improved in the spring sampling session. Some problems did develop with the scales through electrical shortages and scale imbalances, and estimates were recorded at these times.

The weather during January and February was unseasonably mild, which allowed for sorting to proceed unhampered by snow or frozen clumps of garbage. A colder winter would likely yield a different pattern in the waste stream, as yard waste began appearing in February. The amount was small, but did contribute to the overall weight. During both sample periods there was little or no precipitation and all samples were light.

Table 13 Skier Visitation 1988 – 1993

	1988/89	1989/90	1990/91	1991/92	1992/93
Nov	0	0	430	1,825	0
Dec	20,045	28,622	18,544	34,197	3,849
Jan	33,801	42,216	45,915	50,391	17,254

It should be noted that the samples collected in January and February were taken in the first part of the year 1992. The December samples, which were included the winter sampling period, were taken at the end of 1992. Chronologically, this is in the same year, but seasonally they are two different years.

The snowfall in December, 1992, was very low, which resulted in a decrease in skier activity at the Marmot Basin Ski area. In fact, the ski area was not open until Christmas. There was also a significant drop in temperature over the Christmas, season which contributed to low skier visitation. The numbers recorded at the ski hill were as follows:

Obviously, there was a drastic reduction in visitation in 1992/93 from the normal. As a result, restaurants and other food outlets were not open, and the proportion of food in the waste stream also dropped. The effect of weather on the economy is directly reflected in the waste stream, which would not have been so dramatically evident without sampling over this period of time. The drop in temperature (average of -40 degrees celsius over the selected four day sorting period at the end of December and the problem of frostbite) prevented the analysis of garbage over the Christmas period. Three attempts were made to sort garbage at this time, but the conditions were too extreme to accomplish any accurate sorting.

Commercial Waste Winter – 1992

It is difficult to lump all three winter months together as a united whole as several months and changes in climate separated the December sampling period from the January and February sampling period. The trend to decreasing amounts of food in the waste stream can be seen in November, when food waste dropped to 30% from the 40% value in September and October. This downward trend continued into December.

Those categories in the waste stream, which dominated the summer waste stream continued to dominate in the winter, but with some alteration in proportion. The winter months were very dry and caused an alteration in weight between cardboard and the food content of the waste stream.

Both January and February were similar in waste composition and could be treated as a single homogenous interval.

Table 14 Monthly Composition Analysis for Winter – 1992

	Commercial					
Material	**December/92**		**January/91**		**February/91**	
	lbs.	%	lbs.	%	lbs.	%
Food	223.3	25.6	726.4	42.1	712.5	40.0
Yard	0.6	0.0	33.3	1.9		
Wood	2.9	0.3	10.5	0.6	23.9	1.3
Org.	37.8	4.3	56.9	3.4	65.4	3.7
Paper	236.0	27.0	246.4	13.5	256.0	14.4
Cardbd	242.1	27.7	409.3	24.7	409.6	23.0
Plastic	89.2	10.2	131.4	8.3	139.6	7.8
Glass	20.7	2.4	37.8	2.2	41.5	2.3
Metal	19.1	2.2	49.3	3.2	40.1	2.2
Alum.	1.7	0.2	17.9	1.1	20.4	1.1
Min. Mix	20.1	1.2	33.5	1.9		
Min. Oth.	1.1	0.1	6.7	0.4	6.9	0.4
Haz.	0.1	0.0	0.6	0.0		
lbs.	873.9	100.0	1713.4	100.0	1783.3	100.0
Kgs	397.2		778.8		810.6	

Food:

January and February of early 1992 continued to reflect the summer values of 1991, with a slight increase. However, this was not reflected in December when the food value dropped to 28.6%. As mentioned, very few of the commercial outlets were open and those that were, such as Jasper Park Lodge, had a very diminished number of visitors over December, 1992. This was reflected in the diminsished amount of food waste present in their garbage.

Cardboard:

There was a more significant drop in cardboard. In the summer cardboard accounts for 33% of the waste stream, but this dropped to 26% in January and February. Much of this can be attributed to reduction in weight, due to the low humidity and general water content in the cardboard. The December cardboard value remained in line with that of January and February.

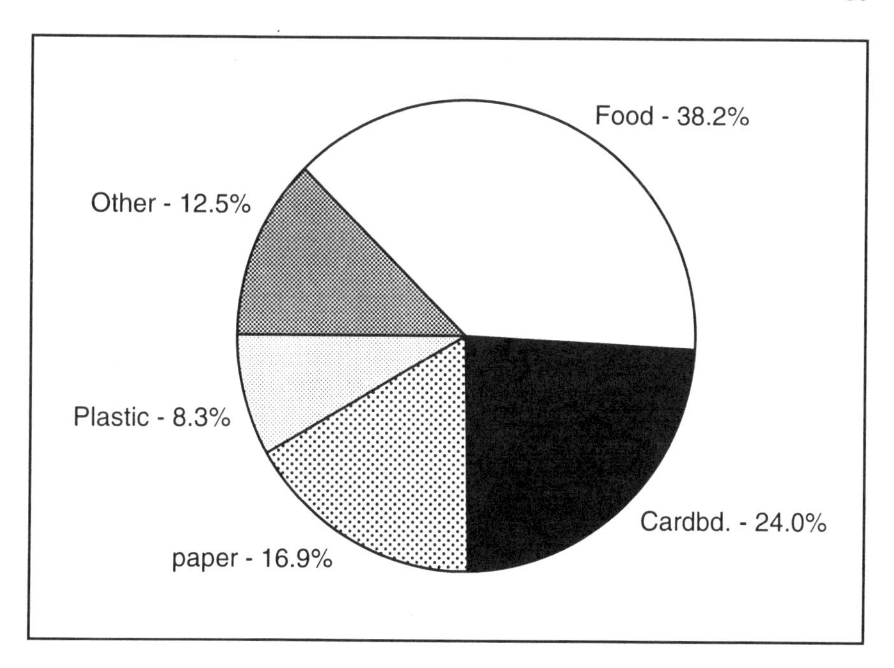

Figure 10 Commercial Composition Winter – 1992

Paper:

Paper increased from 18% in November to 27% in December, which was the highest value recorded for paper out of all twelve months. The December samples were taken at the beginning of the month, when very few outlets were open and the amount of food in the waste stream was down. Most of this paper was newsprint, but there was a substantial increase in paper used in administration of various businesses. There was also an increase in the number of commercial pickups from the post office which, greatly increased the paper content on certain days.

Glass:

Glass decreased by 3.7%, due to the increased use of the bottle depot. Glass is a valuable commodity now being recycled fairly efficiently through this outlet.

Aluminum:

Although aluminum remains a very consistent element in the waste stream, it is also being recycled through the bottle depot. Despite recycling efforts, however, there is no significant change in the amount of this material found in the waste stream from season to season. This may change, as aluminum gives very good financial return for its weight.

Small Elements:

The smaller elements such as wood, aluminum, metal and hazardous waste remained fairly constant in proportion to the summer months. They continue to be low populations in the waste stream and any small variation will reduce the error about the mean at 80% confidence level.

Residential Waste Winter, 1991–1992

More radical changes were observed seasonally in the residential garbage. Because there is no yard waste in the winter which represents a large portion of the summer residential waste stream, other wastes increased proportionally.

Food:

Food waste rose from 14% to 26%. This does not mean that people are more wasteful with food in winter. It merely reflects the change in waste

Table 15 Monthly Composition Analysis for Winter 1991–92

Residential						
Material	**December/92**		**January/91**		**February/91**	
	lbs.	%	lbs.	%	lbs.	%
Food	218.2	25.1	402.0	27.2	377.0	25.7
Yard	3.2	0.4	12.7	0.9	30.7	1.2
Wood	3.7	0.4	17.6	1.2	36.6	1.5
Organic	93.0	10.7	149.5	10.1	248.3	12.7
Paper	171.1	19.7	352.0	23.9	294.0	21.1
Cardbd	158.1	18.2	199.0	13.5	177.2	13.8
Plastic	113.7	13.1	127.6	8.6	145.1	10.0
Glass	66.1	7.6	100.3	6.8	118.5	7.4
Metal	28.8	3.3	65.1	4.4	57.1	3.9
Alum.	7.1	0.8	16.9	1.1	16.3	1.0
Min. Mix	7.6	0.9	24.6	1.7	21.6	1.4
Min. Oth.	5.7	0.4	4.3	0.3		
Haz.	2.3	0.2	1.6	0.1		
lbs.	870.6	100.2	1475.3	100.0	1528.3	100.0
Kg	395.7		670.6		694.7	

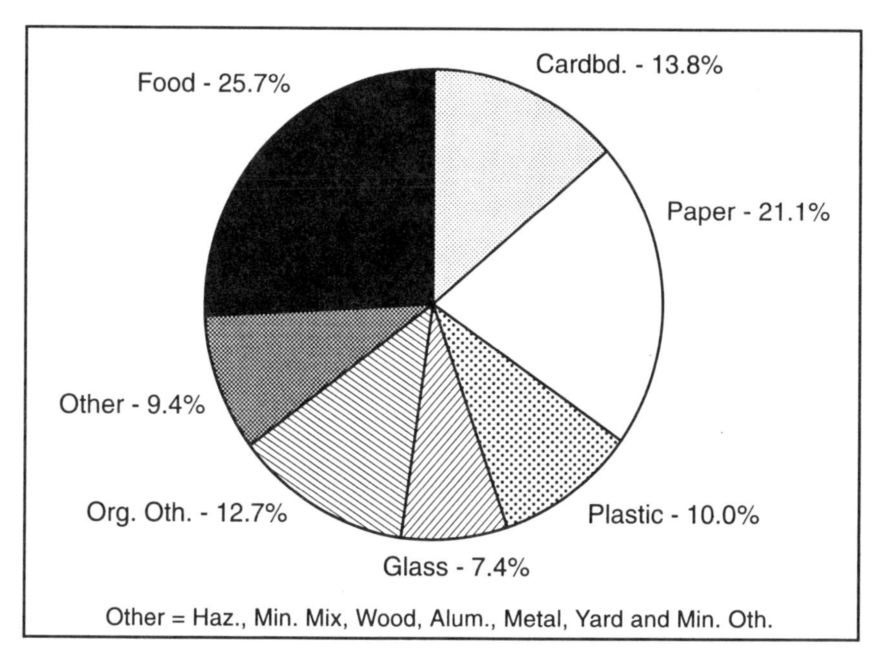

Figure 11 Residential Composition Winter – 1992

composition from summer to winter, through the loss of yard waste.

Other:

Other increases occurred in plastic and organics (textiles); however, cardboard remained fairly constant. The smaller elements of the waste stream also remained unchanged in proportion to the summer months. Hazardous waste did not increase, nor did metals, wood, aluminum or the mineral element (Table 15).

Generation Rates Winter – 1992

The the truck weights recorded in the winter were taken directly from the transfer station scales as apposed to using the scales at scale shed. This was possible because the amount of garbage generated in town had dropped and the number of incoming trucks was low. The trailer could be left out of the building while not being loaded, allowing access to the scales for the incoming trucks. These weights were recorded in kilograms rather than pounds, making final calculation of truck weights to trailer weights simpler. During the winter season only residential and commercial loads were weighed and categorized as both campground and outlying bungalows were closed for the winter. The amount of road-

Table 16 Average Percent Composition by Weight for Each Source Over the Winter Sample Period

Material	Commercial	Residential
Food	38.0%	25.7%
Yard	0.8%	1.2%
Wood	0.9%	1.5%
Organic Oth.	3.7%	12.7%
Paper	16.9%	21.1%
Cardbd.	24.3%	13.8%
Plastic	7.8%	10.0%
Glass	2.3%	7.4%
Metal	2.5%	3.9%
Alum.	0.9%	1.0%
Min. Mix	1.2%	1.4%
Min Oth.	0.3%	0.3%
Haz.	0.0%	0.1%

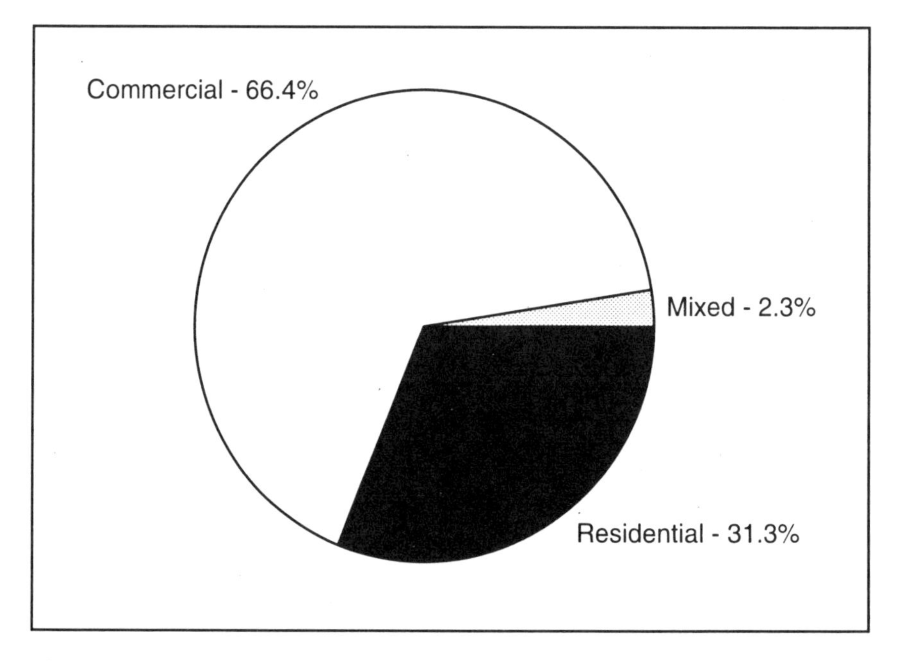

Figure 12 Material Generation Rate Winter – 1992

Table 17 Truck Weights Winter Sample Period

	January	February	December
Scales	51,622 kg	52,240 kg	16,500 kg
	88.8%	88.0%	98.8%
Trailer	58,150 kg	59,370 kg	16,300 kg
	Scales Total:	**103,862 kg**	
	Trailer Total:	**117,520 kg**	
Percent Accuracy:	**91%**		

*NOTE *4 day sample period in December due to very cold weather*

side garbage was weighed on the odd occasion it was picked up, but this was a negligible amount. The weight distribution between commercial and residential changed accordingly. The percent of incoming commercial waste rose to 70%, with the remaining residential waste comprising 31% of the total generated garbage.

As in the summer, all truck weights were calculated over the eight day sample period and compared to that recorded for the transfer trailer over the same period of time. The percent accuracy for both January and February was close to or above 90%. The 10% inaccuracy was due to missing deliveries during the first few days of each shift, when the drivers were reintroduced to the routine of recording truck weights.

On two occasions the scales were inoperative due to mechanical failure or electrical failure. At these times the weights were estimated by the driver. Most experienced drivers were able to accurately assess the weight they were carring and these figures were recorded in lew of actual scale weights.

DATA COLLECTED SPRING – 1992

Spring sampling was conducted at the end of March and into the first week of April. Throughout this time the amount of garbage collected per month increased steadily. In this period there was a steady increase in the amount of yard waste and organic material. The generation rates for each season and monthly amounts are discussed in the sections following dealing with the truck weights taken during each sampling season. In general the total truck weights for the three months increased due to greater volume and increased moisture content. The dry conditions of the winter changed to wetter conditions and higher humidity throughout the spring months. The increase in moisture content, however, was gradual and is not evident in the off seasons. More significant differences can be seen in comparing August with February, where the weather contrast is greatest.

Commercial Waste Spring – 1992

Food:

Food waste continues to dominate the waste stream in the spring months by 36%. This is lower than that found in winter by 5%, but is consistent with the summer months.

Yard:

Yard waste (always low in the commercial waste stream) was the highest in the spring, which would be expected. The percent averages compare in low figures at 4.6% (spring), 1.0% (winter) and 1.4% (summer). That there was any yard waste in the winter at all can only be attributed to

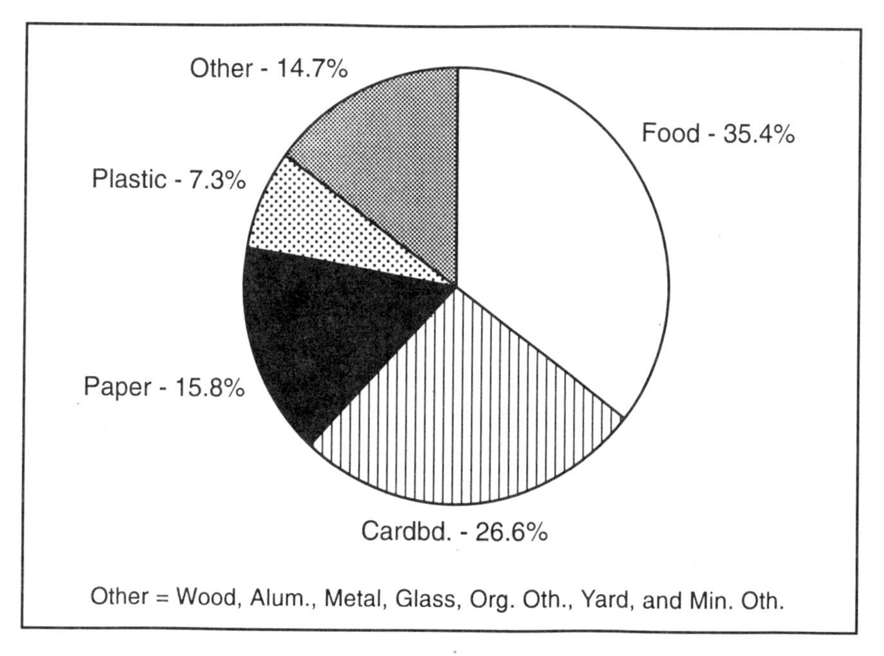

Figure 13 Commercial Composition Spring – 1992

Table 18 Monthly Composition Analysis From Spring 1992

	Commercial					
Material	March		April		May	
	lbs.	%	lbs.	%	lbs.	%
Food	634.7	35.4	760.2	41.5	522.1	30.3
Yard	72.6	4.0	14.1	0.7	156.6	9.1
Wood	1.4	0.1	9.7	0.5	5.1	0.3
Org. Oth.	84.6	4.7	46.5	2.5	66.6	3.9
Paper	296.1	16.5	269.3	14.7	282.2	16.4
Cardbd	440.4	24.6	482.4	26.3	488.3	28.3
Plastic	127.1	7.1	138.8	7.6	122.7	7.1
Glass	34.9	1.9	49.0	2.7	29.1	1.7
Metal	73.7	4.1	30.4	1.7	16.4	0.9
Alum.	21.2	1.2	11.0	0.6	12.8	0.7
Min. Mix	2.5	0.1				
Min. Oth.	3.3	0.2	18.6	1.0	20.3	1.2
Haz.	2.7	0.2	1.1	0.1		
lbs.	1792.7	100.0	1831.1	99.9	1724.7	100.0
Kg	814.9		832.3		784.0	

the unseasonally warm, dry weather in February.

Paper:

Paper waste increased during the spring season by 2% from the winter months and 6% from the summer months. Again, as in winter, much of the paper originated from administration work from all commercial sectors.

Cardboard:

The cardboard figure rose slightly over that in the winter suggesting a seasonal change. Here, moisture content probably plays an increasing role in adding weight to previously dry cardboard.

Plastic:

Plastic waste remained consistent throughout all three seasons (6.6% summer, 8.1% winter and 7.4% spring). Most of the plastic weighed at any time consisted of packaging material rather than returnable bottles.

Organic Other:

The organic category was similar over the winter and spring months. During the summer the figure dropped by about half (1.4% summer, 3.55% winter and 3.6% spring). The minor categories of glass, aluminum, metal, minerals and hazardous waste are very consistent over all the seasons in commercial waste.

Glass:

The amount of glass in the waste stream is less in the spring than the summer. This may be due to increased recycling or the purchase of fewer products in glass containers.

Observations:

An example of the fluctuation in amounts of various materials can be seen in comparing yard waste throughout the three spring months. Yard waste in March was 4% of the waste stream, but dropped to 0.8% in April. It then rose to 9% in May. This coincides with an increase in food waste from 36% in March to 41% in April. The April figures are very similar to the winter months, when yard waste was at its lowest.

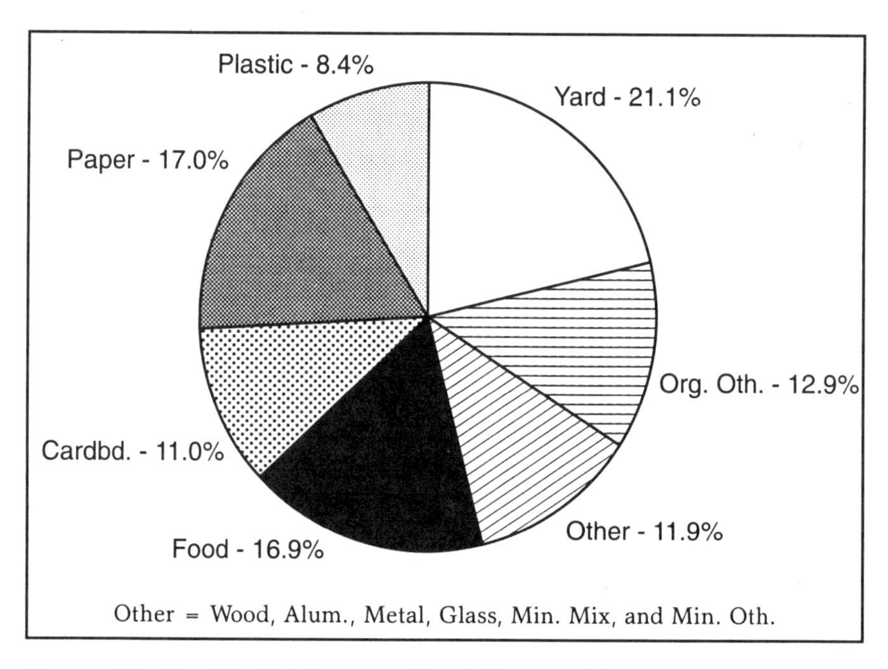

Figure 14 Residential Composition Winter – 1992

Table 19 Monthly Composition Analysis From Spring 1992

Residential						
Material	March		April		May	
	lbs.	%	lbs.	%	lbs.	%
Food	272.2	17.4	278.6	17.6	241.7	15.9
Yard	308.2	19.7	277.2	17.5	403.5	26.5
Wood	24.5	1.6	29.3	1.9	16.4	1.1
Org. Oth.	211.9	13.6	238.7	15.1	147.5	9.7
Paper	265.8	17.0	282.7	17.9	242.4	15.9
Cardbd.	160.6	10.3	181.0	11.4	168.8	11.1
Plastic	121.3	7.8	134.1	8.5	134.0	8.8
Glass	90.6	5.8	55.0	3.5	68.3	4.5
Metal	62.3	4.0	60.0	3.8	52.0	3.4
Alum.	21.4	1.4	17.0	1.1	16.6	1.1
Min. Mix	11.7	0.7	12.6	0.8	1.1	0.1
Min. Oth.	10.5	0.7	12.6	0.8	30.8	2.0
Haz.	2.0	0.1	2.6	0.2	0.5	0.0
lbs.	1563.0	100.1	1581.4	100.1	1523.6	100.1
Kg	710.5		718.8		692.5	

Residential Waste Spring – 1992

The relative distribution and residential waste in the spring compared more favorably with the summer data (Figure 4) than that recorded in the winter (Figure 14). This would be expected with the increase in yard waste as the spring season progressed.

Food:

Food waste is correspondingly lower than that recorded during the winter months by 8.6% and higher than the summer average by 3%.

Yard:

The closer alignment of spring with the summer data is due to the increase in yard waste, which peaks in the spring season in May. Despite the increase, yard waste is still lower by 5% than that in the peak summer season. Yard waste increases from 1.5% in the winter to an average of 21% in the spring. This is an increase of 19.5%.

Organic:

Organic waste, (textiles, etc.), other than yard or food waste, was similar between the winter and spring seasons at 12% and 12.9%, respectively, but above the summer average of 8%. At the time of sorting the waste in March, April and May, organic discards seemed to be at a yearly high, suggesting that residents were house cleaning; however, there must also have been a tendency to discard this material during the winter months as well.

Cardboard:

The amount of cardboard in residential waste is fairly consistent by weight throughout all four seasons, being lowest in the spring and highest in the summer. The average percentage over this period of time is 14% in summer, 16% in fall, 13% winter and 10.8% in the spring. There was also little fluctuation from day to day during each eight day sample period. This consistency indicates that discarded cardboard is not seasonal. As in the summer, most of the cardboard is lightweight packaging material.

Paper:

Paper refuse fluctuates more than cardboard between spring and winter samples. The percentage of paper in the waste stream in the winter was 21.6%, but dropped to 16.5% in the spring. This drop is probably due to the increase in yard waste, which was significant enough to reduce the amount of material picked up in other categories during the winter.

Plastic:

Plastic, like cardboard, remains consistent from season to season. The variation between the four seasons is only 2% between summer and winter. The respective values are 7.76% in the summer, 6.6% in fall, 9.5% in the winter and 7.7% in the spring. All other minor materials remain constant, although, as stated, small fluctuations have a greater impact when the generation rate is low.

Generation Rates Spring – 1992

The lowest waste generation occurred in March, but it steadily increased through April and May (46,750 kg in March, 56,250 kg in April and 97,000 kg in May). Although there were problems with new drivers, the truck weights came much closer to the total weight recorded for the transfer trailer. The breakdown between residential and commercial waste did not change through the first two months, because campground

Table 20 Average Percent Composition By Weight For Each Source Over The Spring Sample Period, 1992

Material	Commercial	Residential
Food	35.8%	17.0%
Yard	4.5%	21.2%
Wood	0.3%	1.5%
Org. Oth.	3.7%	12.8%
Paper	15.8%	16.9%
Cardbd.	26.4%	10.9%
Plastic	7.3%	8.3%
Glass	2.1%	4.6%
Metal	2.3%	3.7%
Alum.	0.8%	1.2%
Min. Mix	0.0%	0.5%
Mix. Oth.	0.8%	1.2%
Haz.	0.1%	0.1%

Table 21 Truck Weights Spring Sample Period

	March	April	May
Scales	53,320 kg	60,170 kg	91,548 kg
	87.7%	93.5%	10.7%
Trailer	46,750 kg	56,250 kg	98,020 kg
	Scales Total:	**205,038 kg**	
	Trailer Total:	**201,020 kg**	
Percent Accuracy: 98.0%			

and outlying motels were still closed. This changed in May, along with a substantial increase in the amount of garbage brought into the transfer station. The truck weights cannot be used for this month; however, because of a change in collection routes, where both commercial and residential garbage were picked up on the same run on several occasions. This problem did not occur in March and April; therefore, the distribution between residential and commercial waste should be accurate. The accuracy of the truck weights, as compared to the transfer trailer weights for this period, also increased. The truck weights in March were 53,320 kg, as compared to 46,750 kg weighed for the transfer trailer (87.7%). This accuracy increased in April to 93.5%, the truck and trailer weights being 60,170 kg versus 56,250 kg, respectively. This accuracy changed in May with the route change and arrival of new drivers. The actual distribution between residential waste and the commercial waste was similar to the winter months with the exception of mixed garbage, which was now picked up in small quantities from outlying highway bins. The breakdown is presented in Figure 15.

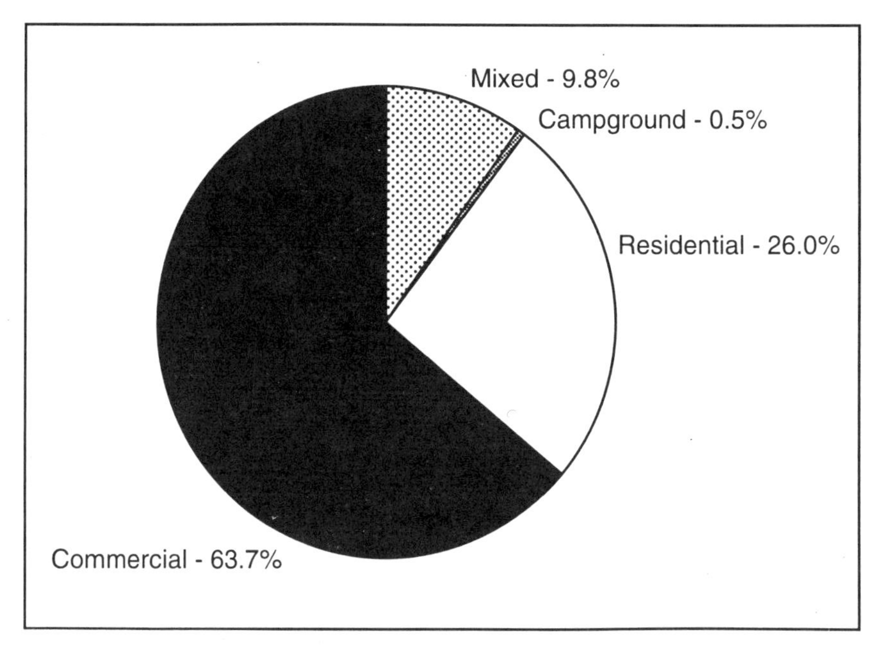

Figure 15 Material Generation Rate Spring – 1992

REVIEW OF TABULATED DATA

As expected, there was considerable seasonal variation in the quantity and distribution of waste. Commercial waste, dominated garbage production throughout the year, but changed proportionately from season to season as other garbage sources such as campground and mixed waste became more predominant in the summer months. The seasonal distribution of waste for Jasper National Park is illustrated graphically in Figure 16.

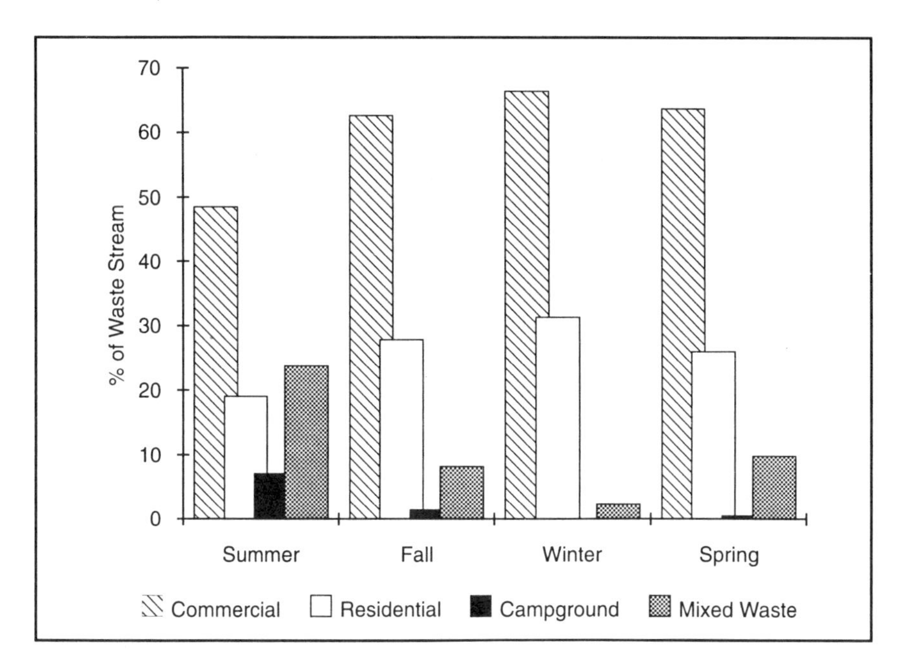

Figure 16 Seasonal Distribution of Waste by Percentage

The seasonal generation rates show a major variation throughout the year. These differences in weight are supported statistically through the null hypothesis, which dicates that there is a difference between summer generation rates and the other three seasons.

The composition of the waste generated seasonally can be compared within sources (i.e., summer commercial food waste as opposed to winter commercial food waste) as well as between sources. The breakdown of the thirteen categories for commercial and residential waste is presented in Table 22.

Within commercial waste, food is seen to be fairly consistent throughout the year, as is plastic and glass. Cardboard, paper and organics other are consistent through fall, winter and spring, but differ in summer; organics and paper are low in the summer while cardboard is high. Yard waste shows the greatest variation, which would be expected from a seasonal product.

Residential food waste increased in the winter, as did paper. However, the increase in paper is not large. This may reflect a more sedentary lifestyle at this time of year and a tendency to purchase more food than is required. The other components are relatively consistent throughout

Table 22 Mean Composition Values

	Summer		Fall		Winter		Spring	
	Comm. %	Res. %	Comm. %	Res. %	Comm. %	Res. %	Comm. %	Res. %
Food	35.5	14.0	37.3	14.3	38.3	25.8	35.4	17.0
Yard	1.3	26.4	0.6	18.6	0.7	1.2	4.6	21.1
Wood	0.9	2.4	0.6	2.0	0.9	1.7	0.3	1.4
Org.Oth.	1.4	8.2	5.0	12.0	3.7	12.3	3.7	12.9
Paper	10.6	15.5	18.7	17.0	17.0	21.0	15.8	17.0
CardBd.	34.1	14.3	24.5	16.1	24.0	13.8	26.6	11.0
Plastic	6.8	6.0	6.6	7.6	8.3	9.9	7.3	8.3
Glass	4.1	5.2	3.1	4.5	2.3	7.5	2.1	4.6
Metal	2.5	4.3	2.1	4.5	2.5	4.1	2.3	3.8
Alum.	1.0	1.4	1.0	1.3	0.9	1.1	0.8	1.2
Min. Mix	0.3	1.6	0.0	1.9	1.3	1.4	0.0	0.6
Min. Oth.	1.3	0.7	0.5	0.4	0.3	0.3	0.8	1.2
Haz.	0.2	0.1	0.0	0.0	0.0	0.1	0.0	0.1

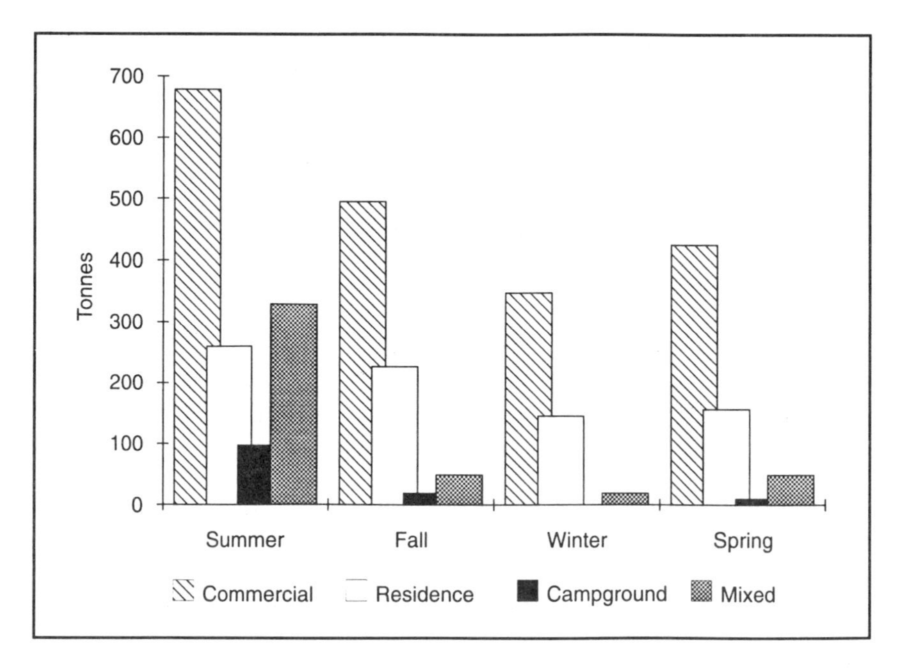

Figure 17 Seasonal Generation Rate Over One Year

the year, with the exception of yard waste, which is, of course, seasonal.

More interesting comparisons can be made between commercial and residential waste. One obvious conclusion is that commercial waste is dominated by food and cardboard waste, while residential waste is more uniform. There is a big difference between the fraction of food, yard, organics and cardboard waste in the commercial waste stream versus the residential waste stream. Plastic, glass and paper are fairly similar in both categories throughout the year.

During each sample session there would be a disproportionate amount of one material over another in a particular category. Any major changes would affect the composition of the whole sample. Since the weight cut off was between 200 and 300 pounds (90 to 136 kg), a large influx of metal or yard material would alter the variation in the remaining categories. Thus, there is a significant variation in an eight day sample period within one source from day to day. Statistics, however, show that the yearly sample size was large enough to dampen these discrepancies and show most categories to obtain a confidence interval within 20%.

One of the more significant influences on the waste stream composition is the weather. Wet weather in the summer caused the garbage to become saturated and therefore heavier. Waste thrown into the garbage bins was subject to contamination from water, not only from rain or snow, but also from other wet garbage. Therefore, the weight of waste before being thrown out can be quite different from what is weighed at the transfer station. This can be considered as generated waste versus received waste. Determination of the water content in the various types of waste can give a closer approximation to what the true weight of a recycled material would be before being discarded. To set up a recycling system for a product based on the terms of measured weight, this could be significant. A prime example of this would be cardboard. Cardboard is extremely porous and can absorb a lot of water, influencing it's true weight in the waste stream. By determining the average water content, a better idea of what should be targeted for recycling can be determined. For any one item this will still vary considerably with the weather. A study done by Bird and Hale (1976) concluded that a general weight loss of 24% between generated and received garbage should be factored in.

An example of water affecting the waste stream can be seen in the summer season, when the apparent weight of cardboard rose significantly in August, causing a percentage loss in other categories. Most of the cardboard was left in open bins and therefore became saturated with water. As cardboard is porous, the water absorption greatly increased its weight.

A second example affecting the waste stream was seen in December, 1992, when a lack of snow in the park caused a decrease in skier visitation and thus a drop in waste production (Table 13). In this instance there was a significant lack of food in the waste stream, which allowed paper and cardboard to appear to be the dominant component.

RESULTS OF STATISTICAL ANALYSIS

The analysis of the data collected over the year was based on the assumption that the data is normally distributed and that it is reasonable to use the Student's test for small samples taken from a large population (see Methods, Section 8.1). A confidence level of 80% was used in the test. Klee (1980) believed the sample data would represent the normal population if the confidence interval fell within 20% of the mean. As the number of samples were collected over the year, the confidence intervals in most categories came within this limit. Certain categories stayed too high, such as the level for hazardous waste, but these materials are a very small fraction of the total waste stream and would require an excessive number of samples to be taken to reduce the confidence interval.

Confidence intervals were computed individually each month for each season, but eight samples were not enough to give a meaningful figure. Therefore, only seasonal data is tabulated in Tables 25 - 28. On the whole the variability in individual categories was less in residential waste than in commercial. This is because there is a more even distribution of waste between categories in residential waste than in commercial waste. The commercial waste stream is composed mainly of food, paper and cardboard, leaving other items a minor percentage of the whole.

The confidence intervals over the entire year, based on the total number of samples for each category, are well within the 20% interval limit in all but the three smaller categories. These materials, however, are only around 1% of the waste stream and are not likely to be targeted for recycling from the waste taken to the transfer station. These yearly confidence intervals are important figures, because they indicate

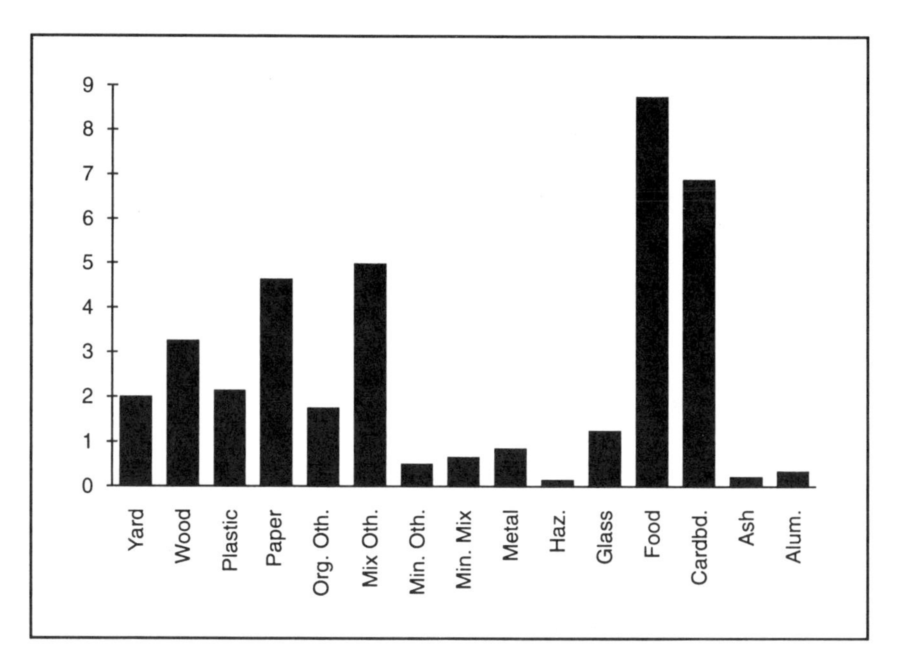

Figure 18 Annual Waste Production

the confidence that can be placed in the total weights calculated for each category over the 12 month sampling period. Table 25 gives the confidence interval for the sample weights, while Table 27, gives the confidence interval for weights over the whole year. These values are higher as they incorporate the sample error plus the truck weight error. Those intervals are still acceptable for categories which represent a large portion of the waste stream.

The intent of this study was to determine the composition of the Jasper waste stream and the amount produced for the individual component on an annual basis. The weight of each component leaving the park annually was calculated from the fraction of the components in the waste stream and the amount of material delivered from each source. The data from the truck weights was converted to percentages on a daily basis and for each season the mean fraction of garbage originating from each source could be estimated.

To determine the fraction of the total waste generated in the park represented by each individual component, it was necessary to combine the means of each component within each source type with the fraction represented by each source. As there was significant variation between each source for both the component fractions and the source fractions, this combination was done on a seasonal basis. The percentage a component represents out of the total waste generated in the Park in a given season can be computed as:

$$T = C * S$$

where T is the percentage that particular component represents of the total garbage generated within its specific source, C is the component value, and S is the percentage of that source, all for the specified season. As both C and S are being estimated from sample data, T will be both an estimated mean and an estimated variance. If the assumption is made that C and S are uncorrelated, and there is no reason to believe otherwise in this case, then the means and variance of T are computed as follows (Benjamin and Cornell, 1970):

$$\text{mean } T = \text{mean } C * \text{mean } S$$

The total weight of waste which leaves the Park varies substantially from season to season, due to the changing number of visitors and transient staff. Unfortunately, it is not possible to average weight figures from year to year seasonally, as the only data on seasonal output is in this study. However, it is possible to estimate the total weight of each component seasonally and annually by using the actual figures for the sample year 1991 - 1992. For each season the weight of a specific component can be computed as:

$$W = T * ts$$

where W is the weight of a particular component, T is the fraction that component represents of the total garbage leaving the park, and 'ts' is the actual weight recorded at the transfer station for that season. In this case 'ts' is considered to be a constant, whereas both T and W are estimates. To compute the mean variance of a constant multiplied by an estimate, the formulas are as follows (Benjamin and Cornell, 1970):

$$\text{mean } W = ts \times \text{mean } T$$
$$\text{var } W = (ts \times ts) \text{ var } T$$

or

mean % component leaving Park per season	×	actual weight leaving Park per season	=	mean total component weight leaving Park per season

These results are presented in Tables 23 and 24. These tables reflect the values for large elements in the waste stream. It should be noted that in each season there was a source that was classified as 'mixed' garbage. This source varied from 23.5% of waste production in the summer to 1.0% in winter. As this source was not sampled at any time, it cannot be included in the above calculations beyond stating that in each season some fraction of the total garbage leaving the park fell into this category.

It is now possible to estimate the total weight for each component over the 1991 - 1992 sampling year. Since samples were taken from three primary sources of garbage it is necessary to combine the data to come up with an estimate of the total overall composition in the park. For a given component, the total weight will be the sum of the components weight in each season for each source. For example, the total weight of food waste will be the weight of food waste from all four seasons added together. As each of these weights are estimates, it is necessary to compute the estimated mean and variance for the total. The mean and variance for a sum of random variables are equivalent to the sum of the means and the sum of the variance respectively (Benjamin and Cornell, 1970). However, to compute a confidence interval around the mean, we must have a *t* value of the selected confidence level, and the value of *t* is always dependent upon the degrees of freedom (n - 1). As *n* (the number of samples) varies with the season and source, the confidence interval is computed for each individual weight described above as W. The sum of these confidence intervals should accurately reflect the interval for the sum of the means. Note that for the weights to add up to the total weight for the year, one must include the weights for mixed garbage and ash, which were accounted for separately during sampling. The annual production of material generation is found in Table 23. These amounts, including ash and mixed garbage, are graphically represented in Figure 20.

For the purposes of recycling programs, it is instructive to know what degree of seasonal variation can be expected. However, if one season is compared to another, the degree of accuracy is limited to the sample size. From examining the confidence intervals for each season, it becomes apparent that only those categories with confidence intervals of less than 20% are worth comparing. One method of determining the relationship between different categories from season to season is to employ the null hypothesis test. Null hypothesis is a statement about a

Table 23 Annual Production of Material Generation (based on items 10% or > in Waste Stream)

	Summer 1991	Fall 1992	Winter 1992	Spring 1992	Total
Food	287,110	217,589	181,014	194,785	880,498
Yard	82,980	44,639	4,688	60,940	193,247
Org. Oth.	34,493	51,314	34,216	41,572	161,595
Paper	125,396	131,425	95,916	104,988	457,725
Cardbd.	273,626	157,517	109,726	143,106	683,975
Plastic	68,181	50,161	46,522	49,241	214,105
Glass	53,202	26,375	20,926	18,843	119,346

population parameter issue (Mason and Douglas, 1988). For recycling purposes the important issue is whether there are significant differences in weight (or generation rate) from season to season. The seasons can be compared as follows:

summer - fall   winter - fall
summer - winter   fall - spring
summer - spring   winter - spring

To determine if one population has a greater weight in a certain category we use a one tailed test for two populations.

The results of this test comparing the seasons as listed are given in Table 27. In all cases, the larger of the two means compared was defined as population 1, and the smaller mean was defined as population 2. From computed t values greater than the critical t value, the alternate hypothesis is that the mean of population 1 is greater than the mean of population 2, at 80% confidence level. An examination of Table 27, indicates that there are greater seasonal differences in commercial waste than in residential waste. Residential waste shows no seasonal difference in several categories, whereas there are several seasonal differences in commercial waste. This is particularly true when summer is compared to the other three seasons. This concurs with personal observations of waste distribution while sorting garbage.

Table 24 Total Number of Kg Generated Per Source Per Season

		Commercial	Residential	Campground
Food	Summer	239,683	35,706	11,721
	Fall	184,947	31,351	1,292
	Winter	137,373	43,642	
	Spring	162,641	31,751	392
Yard	Summer	9,056	67,314	6,611
	Fall	2,989	40,921	729
	Winter	2,706	1,982	
	Spring	21,148	39,571	221
Org. Oth.	Summer	9,517	20,847	4,129
	Fall	24,534	16,325	455
	Winter	13,303	20,914	
	Spring	17,335	24,098	138
Paper	Summer	71,514	39,436	14,446
	Fall	92,598	37,234	35,393
	Winter	60,519	35,397	
	Spring	72,708	31,796	484
Cardbd.	Summer	229,677	36,342	7,607
	Fall	121,285	35,393	839
	Winter	86,331	23,895	
	Spring	122,323	20,529	255
Plastic	Summer	45,791	15,287	7,103
	Fall	32,613	16,765	783
	Winter	29,705	16,818	
	Spring	33,318	15,685	239
Glass	Summer	27,719	13,270	12,213
	Fall	15,214	9,815	1,347
	Winter	6,212	12,714	
	Spring	9,815	8,619	409

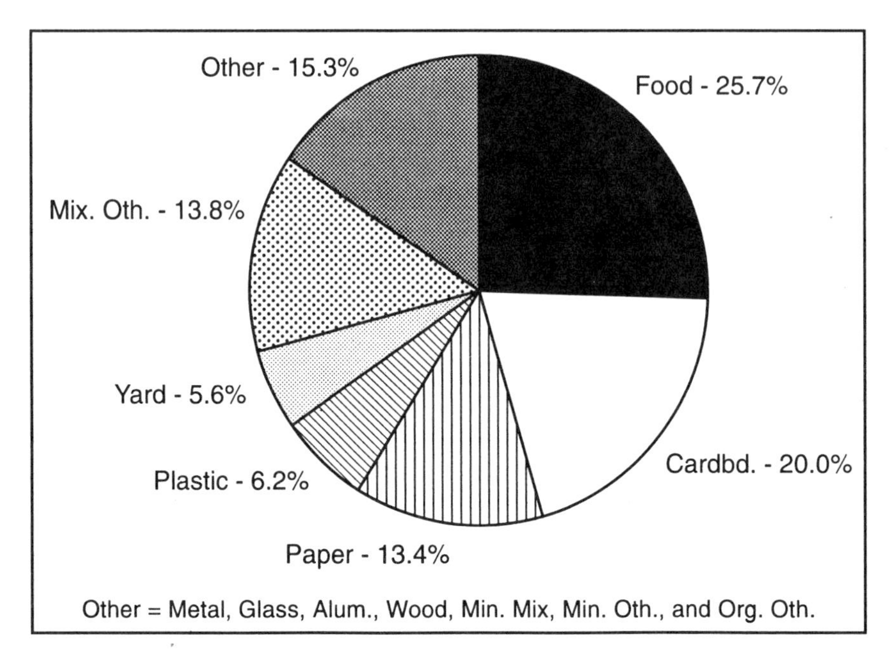

Figure 19 Total Annual Waste By Percentage

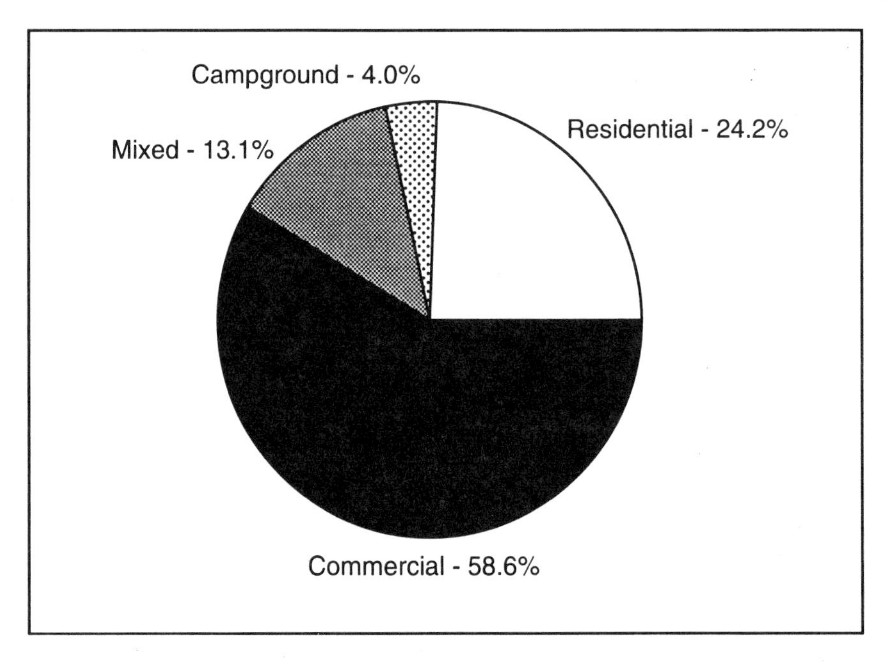

Figure 20 Total Generation Rate Distribution Over One Year

Table 25 Full Year 1991–1992 Data – Statistical Analysis

Commercial (n = 92)

	Food	Yard	Wood	Oth. Org.	Paper	Cardbd.	Plastic
Mean	0.3656	0.0187	0.0067	0.0345	0.1544	0.2744	0.0718
Interval +/-	0.0228	0.0078	0.0015	0.0060	0.0121	0.0183	0.0039
Interval % of Mean	6.24%	41.51%	22.14%	17.52%	7.85%	6.67%	5.41%
Population Std Dev	0.1694	0.0577	0.0111	0.0449	0.0899	0.1358	0.0288
Population Variance	0.0287	0.0033	0.0001	0.0020	0.0081	0.0185	0.0008
Sample Std Dev	0.1684	0.0574	0.0110	0.0446	0.0894	0.1351	0.0286
Sample Variance	0.0284	0.0033	0.0001	0.0020	0.0080	0.0182	0.0008

	Glass	Metal	Alum.	Mineral	Mix Min.	Oth.	Haz.
Mean	0.0293	0.0232	0.0093	0.0037	0.0074	0.0009	
Interval +/-	0.0034	0.0031	0.0011	0.0020	0.0030	0.0005	
Interval % of Mean	11.71%	13.44%	11.52%	54.81%	40.47%	51.71%	
Population Std Dev	0.0255	0.0232	0.0080	0.0152	0.0221	0.0034	
Population Variance	0.0006	0.0005	0.0001	0.0002	0.0005	0.0000	
Sample Std Dev	0.0253	0.0230	0.0079	0.0151	0.0220	0.0033	
Sample Variance	0.0006	0.0005	0.0001	0.0002	0.0005	0.0000	

Residential (n = 83)

	Food	Yard	Wood	Oth. Org.	Paper	Cardbd.	Plastic
Mean	0.1737	0.1763	0.0190	0.1123	0.1740	0.1380	0.0786
Interval +/-	0.0109	0.0239	0.0042	0.0105	0.0091	0.0073	0.0047
Interval % of Mean	6.28%	13.56%	22.04%	9.31%	5.25%	5.30%	6.04%
Population Std Dev	0.0768	0.1683	0.0295	0.0737	0.0644	0.0515	0.0335
Population Variance	0.0059	0.0283	0.0009	0.0054	0.0041	0.0027	0.0011
Sample Std Dev	0.0763	0.1673	0.0293	0.0732	0.0640	0.0512	0.0333
Sample Variance	0.0058	0.0280	0.0009	0.0054	0.0041	0.0026	0.0011

	Glass	Metal	Alum.	Mineral	Mix Min.	Oth.	Haz.
Mean	0.0537	0.0416	0.0123	0.0131	0.0065	0.0008	
Interval +/-	0.0042	0.0036	0.0011	0.0048	0.0019	0.0003	
Interval % of Mean	7.92%	8.72%	8.92%	36.61%	29.85%	33.59%	
Population Std Dev	0.0299	0.0256	0.0077	0.0337	0.0137	0.0020	
Population Variance	0.0009	0.0007	0.0001	0.0011	0.0002	0.0000	
Sample Std Dev	0.0297	0.0254	0.0077	0.0335	0.0136	0.0020	
Sample Variance	0.0009	0.0006	0.0001	0.0011	0.0002	0.0000	

* Interval +/- and Interval percent of mean are computed at 80% confidence level. Commercial t = 1.2923; Residential t = 1.2934.

Table 25 (continued)

Campground (n = 36)

	Food	Yard	Wood	Oth. Org.	Paper	Cardbd.	Plastic
Mean	0.1188	0.0670	0.0002	0.0419	0.1464	0.0771	0.0720
Interval +/-	0.0158	0.0291	0.0002	0.0070	0.0123	0.0090	0.0074
Interval % of Mean	13.29%	43.36%	96.54%	16.67%	8.38%	11.65%	10.21%
Population Std Dev	0.0725	0.1334	0.0010	0.0320	0.0564	0.0412	0.0337
Population Variance	0.0053	0.0178	0.0000	0.0010	0.0032	0.0017	0.0011
Sample Std Dev	0.0715	0.1315	0.0010	0.0316	0.0556	0.0407	0.0333
Sample Variance	0.0051	0.0173	0.0000	0.0010	0.0031	0.0017	0.0011

	Glass	Metal	Alum.	Mineral	Mix Min.	Oth.	Haz.
Mean	0.1238	0.0409	0.0253	0.2798	0.0052	0.0016	
Interval +/-	0.0126	0.0056	0.0028	0.0372	0.0026	0.0008	
Interval % of Mean	10.17%	13.72%	11.28%	13.28%	50.98%	47.05%	
Population Std Dev	0.0578	0.0258	0.0131	0.1707	0.0121	0.0035	
Population Variance	0.0033	0.0007	0.0002	0.0291	0.0001	0.0000	
Sample Std Dev	0.0570	0.0254	0.0129	0.1683	0.0119	0.0034	
Sample Variance	0.0032	0.0006	0.0002	0.0283	0.0001	0.0000	

* Interval +/- and Interval percent of mean are computed at 80% confidence level. Campground t = 1.3069.

Table 26 Total Weight For Each Category And Source By Season (kg)

Commercial		Food	Yard	Wood	Oth.Org.	Paper	Cardbd.	Plastic
Summer 91								
Mean		239,683	9,056	6,383	9,517	71,514	229,677	45,791
n = 24	Samp Var	1.37E10	4.75E8	8.92E7	6.71E7	1.58E9	1.22E10	6.11E8
t = 1.3199	interval	31,490	5,873	2,544	2,208	10,721	29,788	6,661
Fall 92								
Mean		184,947	2,989	3,123	24,534	92,598	121,285	32,613
n = 24	Samp Var	1.04E10	2.72E7	2.75E7	1.07E9	3.01E9	3.68E9	2.39E8
t = 1.3199	interval	27,523	1,405	1,412	8,810	14,782	16,353	4,169
Winter 92								
Mean		137,373	2,706	3,038	13,303	60,519	86,331	29,705
n = 20	Samp Var	3.79E9	1.13E8	2.02E7	1.76E8	1.58E9	2.63E9	1.77E8
t = 1.3291	interval	18,306	3,154	1,335	3,947	11,820	15,229	3,951
Spring 92								
Mean		162,641	21,148	1,385	17,335	72,708	122,323	33,318
n = 24	Samp Var	8.74E9	2.22E9	6.91E6	3.97E8	1.44E9	4.84E9	1.38E8
t = 1.3199	interval	25,188	12,706	708	5,371	10,212	18,750	3,163
		Glass	**Metal**	**Alum.**	**Min.Mix**	**Min.Oth.**	**Haz.**	
Summer 92								
Mean		27,719	16,586	6,597	2,313	8,494	1,234	
n = 24	Samp Var	6.04E8	3.01E8	5.03E7	4.38E7	6.22E8	1.35E7	
t = 1.3199	interval	6,621	4,678	1,910	1,783	6,719	990	
Fall 92								
Mean		15,214	10,445	4,948	0	2,364	359	
n = 24	Samp Var	8.48E7	9.35E7	1.11E7	0.00	5.77E7	2.77E6	
t = 1.3199	interval	2,480	2,605	897	0	2,046	448	
Winter 92								
Mean		8,212	8,932	3,283	4,473	1,219	54	
n = 20	Samp Var	1.49E7	4.03E7	7.20E6	1.12E8	7.65E6	3.71E4	
t = 1.3291	interval	1,149	1,886	797	3,143	822	57	
Spring 92								
Mean		9,815	10,359	3,860	222	3,690	308	
n = 24	Samp Var	1.56E8	1.96E8	1.40E7	1.21E6	5.58E7	4.91E5	
t = 1.3199	interval	3,369	3,773	1,008	296	2,012	189	

Table 26 (continued)

Residential		Food	Yard	Wood	Oth.Org.	Paper	Cardbd.	Plastic
Summer 91								
Mean		35,706	67,314	6,136	20,847	39,436	36,342	15,287
n=23	Samp Var	7.28E8	3.90E9	1.27E8	4.46E8	9.40E8	6.90E8	1.27E8
t=1.3216	interval	7,437	17,214	3,104	5,820	8,451	7,240	3,100
Fall 92								
Mean		31,351	40,921	4,369	26,325	37,234	35,393	16,765
n=21	Samp Var	5.62E8	1.46E9	6.17E7	3.85E8	5.25E8	4.34E8	1.33E8
t=1.3266	interval	6,864	11,064	2,274	5,680	6,631	6,028	3,340
Winter 92								
Mean		43,642	1,982	2,853	20,914	35,397	23,395	16,818
n=18	Samp Var	4.22E8	2.57E7	2.54E7	2.67E8	3.59E8	1.58E8	7.29E7
t=1.3348	interval	6,463	1,594	1,586	5,143	5,961	3,960	2,686
Spring 92								
Mean		31,751	39,571	2,703	24,098	31,796	20,529	15,685
n=21	Samp Var	3.60E8	1.50E9	1.34E7	4.63E8	3.27E8	1.66E8	1.14E8
t=1.3266	interval	5,491	11,214	1,060	6,230	5,233	3,731	3,089
		Glass	**Metal**	**Alum.**	**Min. Mix**	**Min. Oth.**	**Haz.**	
Summer 91								
Mean		13,270	11,030	3,509	3,692	1799	301	
n=23	Samp Var	1.37E8	1.53E8	1.07E7	7.54E7	1.42E7	5.45E5	
t=1.3216	interval	3,227	3,409	902	2,392	1,040	203	
Fall 92								
Mean		9,815	9,796	2,809	4,082	965	28	
n=21	Samp Var	5.39E7	4.40E7	7.63E6	1.89E8	5.48E6	1.51E4	
t=1.3266	interval	2,125	1,921	800	3,980	678	36	
Winter 92								
Mean		12,714	6,909	1,786	2,301	420	164	
n=18	Samp Var	5.40E7	2.04E7	1.11E6	1.12E7	6.23E5	1.79E5	
t=1.3348	interval	2,312	1,422	331	1,055	248	133	
Spring 92								
Mean		8,619	7,043	2,190	1,040	2.147	195	
n=21	Samp Var	4.74E7	2.67E7	2.33E6	5.09E6	1.91E7	1.39E5	
t=1.3266	interval	1,993	1,495	442	653	1,267	108	

Table 26 (continued)

Campground		Food	Yard	Wood	Oth.Org.	Paper	Cardbd.	Plastic
Summer 92								
Mean		11,721	6,611	22	4,129	14,446	7,607	7,103
n=24	Samp Var	1.13E8	2.40E8	1.31E4	1.88E7	1.11E8	4.11E7	3.15E7
t=1.3199	interval	2,865	4,176	31	1,167	2,837	1,728	1,512
Fall 92								
Mean		1,292	729	2	455	1,593	839	783
n=24	Samp Var	1.31E7	1.62E7	7.96E2	1.91E6	1.63E7	5.14E6	4.23E6
t=1.3199	interval	976	1,085	8	372	1,089	611	554
Spring 92								
Mean		392	221	1	138	484	255	238
n=24	Samp Var	4.88E6	5.66E6	2.72E2	7.01E5	6.19E6	1.92E6	1.59E6
t=1.3199	interval	595	641	4	226	670	374	340
		Glass	**Metal**	**Alum.**	**Min.Mix**	**Min.Oth.**	**Haz.**	
Summer 91								
Mean		12,213	4,035	2,494	27,608	511	158	
n=24	Samp Var	9.28E7	1.39E7	4.27E6	6.27E8	1.95E6	1.60E5	
t=1.3199	interval	2,596	1,005	557	6,746	376	108	
Fall 92								
Mean		1,347	445	275	3,044	56	17	
n=24	Samp Var	1.25E7	1.59E6	5.44E5	7.28E7	1.27E5	1.06E4	
t=1.3199	interval	952	339	199	2,298	96	28	
Spring 92								
Mean		409	135	83	924	17	5	
n=24	Samp Var	4.70E6	5.89E5	2.04E5	2.71E7	4.41E4	3.70E3	
t=1.3199	interval	584	207	122	1,402	57	16	

Table 27 Comparisons of Mean Weight Between Seasons Computed t-value

	Commercial						
	Food	Yard	Wood	Oth.Org.	Paper	Cardbd.	Plastic
tcrit (80%)							
0.850							
sum-fall	1.727	1.326	1.478	2.182	1.524	4.210	2.213
0.850							
sum-win	3.606	1.217	1.485	1.185	0.934	5.458	2.670
0.850							
sum-spg	2.522	1.140	2.498	1.777	0.106	4.025	2.232
0.850							
fall-win	1.823	0.115	0.057	1.438	2.179	2.039	0.661
0.850							
fall-spg	0.789	1.875	1.452	0.921	1.461	0.055	0.178
0.850							
win-spg	1.035	1.710	1.520	0.772	1.039	1.918	0.957
	Glass	Metal	Alum.	Min.Mix	Min.Oth.	Haz.	
tcrit (80%)							
0.850							
sum-fall	2.334	1.514	1.031	1.712	1.152	1.062	
0.850							
sum-win	3.589	1.911	2.019	0.846	1.326	1.465	
0.850							
sum-spg	3.181	1.367	1.672	1.527	0.904	1.213	
0.850							
fall-win	3.171	0.600	1.801	2.077	0.638	0.814	
0.850							
fall-spg	1.703	0.025	1.064	0.989	0.610	0.139	
0.850							
win-spg	0.551	0.420	0.578	1.961	1.400	1.570	

Table 27 (continued)

Residential

	tcrit (80%)	**Food**	**Yard**	**Wood**	**Oth.Org.**	**Paper**	**Cardbd.**	**Plastic**
sum-fall	0.850	0.566	1.671	0.598	0.889	0.268	0.132	0.430
sum-win	0.851	1.034	4.414	1.147	0.011	0.490	1.922	0.479
sum-spg	0.850	0.557	1.750	1.333	0.505	0.994	2.496	0.120
fall-win	0.852	1.715	4.282	0.704	0.926	0.270	2.131	0.016
fall-spg	0.851	0.060	0.114	0.881	0.350	0.854	2.782	0.315
win-spg	0.852	1.879	4.080	0.107	0.513	0.607	0.700	0.362

	tcrit (80%)	**Glass**	**Metal**	**Alum.**	**Min.Mix**	**Min.Oth.**	**Haz.**
sum-fall	0.850	1.159	0.406	0.763	0.114	0.871	1.672
sum-win	0.851	0.176	1.342	2.143	0.642	1.520	0.704
sum-spg	0.850	1.586	1.371	1.685	1.357	0.283	0.591
fall-win	0.852	1.229	1.560	1.479	0.535	0.941	1.403
fall-spg	0.851	0.545	1.500	0.898	1.000	1.091	1.952
win-spg	0.852	1.795	0.086	0.947	1.394	1.648	0.249

* Critical t-values are for a one-sided test of whether one mean is significantly greater than the other. In all cases, the larger of the two means compared was defined as population 1, and the smaller mean was defined as population 2. For computed t values greater than the critical t-value, we accept the alternate hypothesis that the mean of population 1 is greater than the mean of population 2 at an 80% level of confidence.

Table 28 Total Weights From All Sources by Season

	Food	Yard	Wood	Oth.Org.	Paper	Cardbd.	Plastic
Summer 91							
mean	287,110	82,980	12,541	34,493	125,396	273,626	68,181
samp var	1.45E10	4.62E9	2.16E8	5.32E8	2.63E9	1.30E10	7.69E8
interval	41,791	27,263	5,680	9,195	22,009	38,755	11,273
Fall 92							
mean	217,589	44,639	7,495	51,314	131,425	157,517	50,161
samp var	1.10E10	1.50E9	8.92E7	1.46E9	3.55E9	4.12E9	3.77E8
interval	35,363	13,555	3,693	14,862	22,502	22,992	8,063
Winter 92							
mean	181,014	4,688	5,891	34,216	95,916	109,726	46,522
samp var	4.22E9	1.38E8	4.56E7	4.44E8	1.94E9	2.78E9	2.50E8
interval	24,769	4,748	2,921	9,091	17,781	19,190	6,637
Spring 92							
mean	194,785	60,940	4,088	41,572	104,988	143,106	49,241
samp var	9.11E9	3.73E9	2.03E7	8.61E8	1.77E9	5.01E9	2.53E8
interval	31,274	24,561	1,773	11,827	16,115	22,855	6,592
Grand Total							
mean	880,498	193,246	30,015	161,596	457,726	683,975	214,105
samp var	3.88E10	9.99E9	3.71E8	3.29E9	9.90E9	2.49E10	1.65E9
interval	133,197	70,126	14,066	44,975	78,407	103,792	32,565
% interval	15.13%	36.29%	46.87%	27.83%	17.13%	15.17%	15.21%

Table 28 (continued)

	Total	Glass	Metal	Alum.	Min, Mix	Min. Oth.Haz.
Summer 91						
mean	53,202	31,651	12,599	33,613	10,804	1,693
samp var	8.34E8	4.68E8	6.53E7	7.46E8	6.38E8	1.42E7
interval	12,444	9,092	3,370	10,921	8,135	1,301
Fall 92						
mean	26,375	20,686	8,032	7,126	3,385	405
samp var	1.51E8	1.39E8	1.93E7	2.62E8	6.33E7	2.79E6
interval	5,557	4,865	1,896	6,278	2,820	512
Winter 92						
mean	20,926	15,841	5,069	6,774	1,640	218
samp var	6.89E7	6.07E7	8.31E6	1.23E8	8.27E6	2.16E5
interval	3,461	3,308	1,129	4,198	1,070	190
Spring 92						
mean	18,843	17,537	6,134	2,187	5,854	509
samp var	2.08E8	2.23E8	1.65E7	3.34E7	7.49E7	6.34E5
interval	5,946	5,475	1,571	2,351	3,335	313
Grand Total						
mean	119,346	85,715	31,834	49,699	21,682	2,824
samp var	1.26E9	8.92E8	1.09E8	1.16E9	7.85E8	1.78E7
interval	27,408	22,740	7,965	23,748	15,360	2,316
% interval	22.96%	26.53%	25.02%	47.78%	70.84%	82.00%

6 Transportation

GENERAL OBSERVATION

Waste, once generated, must be collected and transported to a waste disposal facility. At present, waste in North America is being produced in excess of 2.25 kg per person per day (Gore, 1991). Disposal of this waste is not an easy solution, as it requires land which can accommodate hundreds of tonnes of garbage per year. Jasper currently faces this problem with trade waste. If waste is not disposed of within the park it must be transported to another location, as is the case with MSW. In reviewing the problems and cost of handling all waste material generated in the Park, it becomes apparent that transportation is a major part of waste management. For this reason it is important to examine the issue of transportation and how it relates to the Park.

One reason why transportation issues, as a part of the waste management, are becoming important is the problem of locating landfill sites that are suitable for waste disposal. This may mean relocation to an area with a dry climate. Economically, much of the land surrounding urban centres is too costly for a community to purchase the required acreage needed for a landfill site, further promoting the moving of this facility to cheaper land (Anon. Town Looks Ahead, Buys Baler, 1978). Permanent disposal facilities are also being located far from the generation source due to the "not in my backyard" attitude of people who live in the area cited for a landfill or incinerator (Howlett, 1990). For these reasons both rural and urban communities must face increasing transportation costs and associated problems. This does not differ from transportation problems in general, in that a product must be moved from its source to its disposal site. In loading and unloading these goods, handling costs are generated which must be accounted for by either the transport company or the producer. These costs are a part of the transportation process. The process may be broken down as follows:

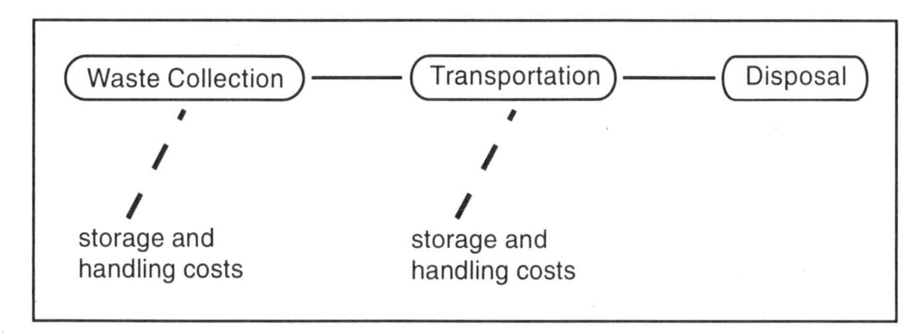

Figure 21 The Waste Disposal Process

While all parts of the solid waste management system are important some portions have received more attention then others. One critical factor is how waste is generated, as the quantity, composition and seasonal variation determines the method of collection, storage, transportation, and disposal (Reindle, 1977).

Waste Collection

Collection of waste includes all processes involved in moving that material to a collection depot (National Waste Reduction Handbook, 1991). Waste collection can be a costly procedure if the methods of transport and route selection are not carefully considered. The collection process has been estimated to account For up to 75% to 80% of the amount being spent on transporting MSW in North America (Cardile and Verhoff, 1974). Although every community is unique, efficient refuse collection requires that equipment performance standards be tailored to the local conditions (SCS Engineers, 1979). For this reason, several communities in the United States are re-evaluating their collection and delivery methods with regard to cost effectiveness (Cardile and Verhoff, 1974). Truck size was evaluated in South Bend, Indiana, to minimize costs. Time and motion data were obtained on three types of collection systems operating in the area (Cardile and Verhoff, 1974). The most economical truck size was calculated for each collection district and for several disposal sites.

In Tulsa, Oklahoma, a waste management committee was formed by the local citizens to oversee collection practices and costs. They opted to switch from gasoline powered trucks to diesel, which resulted in a 30% savings in fuel and substantially reduced maintenance requirements (Tulsa Refuse Department Competes for Routes, Employees And Landfill space, 1979). In Boston, Mass., curbside rubbish collection is a problem because the city is characterized by steep hills and narrow streets and alleys. Boston now employes a private trucking firm, which reduced many of these problems by using small trucks and well trained efficient drivers. By using hand picked drivers and trucks equipped with loading cranes, the number of trips to the land fill was reduced from 7 loads per day to 3, leading to an operating efficiency of 94%. In some cases privatization of pickup services has resulted in significant savings. Systems Disposal Service, a private collection firm, replaced the municipal collection service in Huntington Park, California. The decision to do this came from financial cutbacks, increasing insurance, overhead and depreciation costs. The new service replaced the city's nine workers, who spent 50% of their time collecting refuse, with four of their own workers. They also switched to step-inside loaders, which can be operated by one person and can haul 11 to 12 tons per load. This cuts in half the number of trips to the landfill and accounts for a large portion of the systems efficiency. The city estimates a saving of $130,000 in the first year of operation (Anon, Private Hauler Accepts California City Proposition, 1979).

The material collected may be taken directly to a disposal facility such as a landfill or local incinerator, or more commonly to a collection depot. Collection depots are either transfer stations, or a recycling centre. With the development of large regional landfills, transfer stations will become much more common (Woods, 1991). One problem encountered with the transportation of waste is the storage of the refuse until it can be transported to the final destination, ie. landfill, incinerator, or recycling center (Howlett, 1990). For truck or rail hauling to be economical, the rail car or truck must maximize its carrying capacity. If the storage area in a depot is small, the operator may ask a transport company to leave the vehicle on site until it can be filled. Similarly, if the depot to which it is being transported is small, it may take one or more days to unload it (pers. comm., Siedel, 1992). This will add considerably to the transportation costs of moving that material. This problem crops up mainly with recycled goods, as transfer stations are large enough to contain the carrying capacity of a transfer trailer. Regional landfills should be designed to meet this capacity and no delay should occur in disposing of the waste (pers. comm. Polski, 1993). This is a particular problem in small rural communities that use improvised centers or that improperly plan the size of the depot they build, either due to financial constraints or unrealistic estimates of the volume of the material received (Siedel, 1992).

Transportation of Waste to Disposal Area

Once the waste has been collected, it may be transferred to the disposal area by either truck or rail. The decision to transport the waste to a transfer station or directly to the dump site will depend on the difficulty of transferring the waste to the selected mode of transport and the cost involved. For short hauls, trucks are most often used, as the material can be directly loaded and unloaded either by hand or forklift (recycled goods) or front end loader (MSW) thereby reducing handling costs. Trucks are also more flexible in destination sites and are usually the only option available (pers. comm., Oberg, 1993). Truck transportation is less expensive than rail, but, is limited in carrying capacity. The average trucking cost is $30 to $35 per tonne for every trip to Vancouver (1,225 km). This works out to 2.9¢ a tonne per kilometre, compared with $40 per tonne or 3.3¢ a tonne per km, for rail transport (pers. comm. Orydzuk, 1993). The limited carrying capacity presents a problem in using truck transport if volumes are high and distances long. Paperboard Industries in Edmonton truck 125 to 140 tonnes to Vancouver daily, but, must use 25 different trucking companies to accomplish this. They find a great deal of time is taken up in contracting and arranging this transfer and would prefer to go by rail if they could receive competitive prices. This would mean dealing with only one source, saving time and money (pers. comm., Orydzuk, 1993).

In the United States rail transport is reported to be cheaper than Canadian rates and rail haul is being used, increasingly to transport waste (Fergueson, 1978). One railroad source indicated that it costs about $1.40 a km to move a 75 ton boxcar of waste or 1.9 cents a tonne per kilometre (Barns, 1974). This cost can vary in different areas. The 1993 prices quoted for Seattle, Wash. are $20 (Can.) per tonne (American figures converted to Canadian figures). The haul distances is approximately 500 km, which makes this 4.0 cents per tonne per kilometre. This rate is 3.2 cents per tonne per km in U.S. currency. These costs do not include truck transport to the railhead or disposal site. The distances involved do not vary in price once the MSW is loaded on the rail car, as the expenses do not increase over distance with rail haul. There was an estimate that the rate would increase by a cent or two over shorter hauls (pers. comm., Jones, 1993).

The use of railroads in hauling MSW is not new in the United States; It has increased from 6.2 million tonnes in 1988 to 7.2 million tonnes in 1990

(Woods, 1991). In the past decade with landfills closing in the Pacific Northwest and West Coast, rail haul of MSW to less expensive landfills inland, has emerged as a viable long term solution to disposal (Woods, 1991). The best example of rail haul in the northwest is the city of Seattle, Wash. As of April 1991, Seattle became the first known high volume, long distance rail haul operation of MSW in the United States (Hughes, 1992). This is not limited to the Pacific Northwest. The city of Atlanta, Georgia, has invested in a baling system to allow MSW to be hauled by rail to alternate landfill locations (Barns, 1974). For long range transportation of MSW, rail haul has the advantage of moving large loads over long distances. For these communities there is a cost saving inherent in the reduction of the number of trips by truck to the landfill as well as a reduction in pollution and highway traffic problems (Woods, 1991).

Rail haul is used only in very large centres where the MSW exceeds 500,000 tonnes per year. Smaller centres still make use of local landfills if it is within 60 kilometres of the waste generation center. Although there is no recorded truck to rail ratio, truck hauls are still the predominate mode of transport of MSW (pers. comm., Jones, 1993).

Waste Disposal

Disposal of waste is part of the removal of MSW from a given environment and must be included in the overall cost of waste transportation. In Canada the majority of waste is placed in landfills (Howlett, 1990). Trucks bringing waste to such a facility may be weighed before depositing the waste in the dump. Once the waste is dumped, it must be buried and then covered with fresh fill to prevent water accumulation, wind, erosion and animal contamination. This also prevents excessive odor (O'leary and Walsh, 1991). Most landfill operators use bulldozers to dig trenches to bury the waste and excavate unused areas for fill to cover the dumped material.

The tipping fees (cost per tonne to dispose of one tonne of waste at a landfill) are supposed to cover these costs and are generally quoted as the disposal cost. Most of these fees, however, do not reflect the true cost of disposal. Expenses, such as rehabilitation of the closed landill and development of a new landfill are not usually considered (Howlett, 1990). These costs are very specific to the location of the community and have to be assessed locally.

COLLECTION, TRANSPORTATION AND DISPOSAL OF WASTE IN JASPER NATIONAL PARK

Trade waste is all waste created by construction, demolition, excavation and discards from business and households, that cannot be construed as municapal solid waste (*Recycling of Waste in Alberta, a Technical Report*, 1987).

Jasper generates waste much like any community, with the exception that most of this waste is produced in the summer months in response to the tourist influx. Although the waste is seasonal, disposal problems remain the same throughout the year. Jasper's waste can be classified as either trade waste, municipal solid waste or recycled waste. The three types of waste are different in substance and source, and consequently the disposal of this waste is also different. Municipal solid waste and recycled waste is disposed of outside the park boundaries. The destination of this waste can best be illustrated in a flow chart, (see Figure 21).

The main issues for park management is one of cost. An important question is: would the park be wiser to privatize all transportation or try to ecomonize on their own operation? Another point is the viability of switching to recycling as a way of disposing of our waste (*National Round Table on the Environment & Economy*). The material is still being transported, but, to a different disposal site. There are diversionary credits to be obtained by saving landfill costs, but these costs are difficult to calculate.

Disposal in the Park

Trade Waste

Trade waste is the only material that is disposed of within the park. Over half of this material is brought in by private contractors hauling demolition and construction debris from sites within the townsite. The trucks hauling this waste are generally tandem trucks, capable of carrying 5,500 kg of material (pers. comm., Lonsberry, 1993). There are seven companies currently trucking in Jasper.

The second major source of trade waste comes from Parks Canada. This waste is mostly debris generated through government projects, clean up and maintenance. A lot of this material is old or dirty fill from former building sites or sewer lines. Other contributers are CN Rail, hotels with their own disposal vehicles, Alberta Power and private households (pers.

comm., Greer, 1993). The distance involved in hauling waste to the trade waste pit depends on the source; however, areas that generate trade waste are limited and may be listed in order of primary generation (see Figure 23):

Jasper townsite. (6 km)
Jasper Tramway and Whistler's Hostel. (12 km)
Marmot Basin Ski Area. (30 km)
The Icefields Parkway and Sunwapta Bungalows. (120 km)
Snaring road and campsite. (10 km)
Pocahontas/Miette area east of Jasper. (45 km)
Maligne Lake. (45 km)

As with MSW, trade waste is seasonal. The generation of construction waste seasonally, in Jasper would not likely be very different from that in any northern community where construction occurs during late spring, summer and early fall.

Disposal Outside the Park

Municipal Solid Waste

Jasper's MSW as mentioned, is disposed of at the Hinton Regional Landfill and is transported there in the park transfer trailer. MSW is generated primarily within the townsite, however, when considering transportation, generation of waste from outlying areas must be considered. The drivers break up the park into three distinct collection routes:

1.) The South Run - the haul from the Icefields Parkway where the Icefield Chalet, Sunwapta Bungalows, five small campgrounds and wayside bins are picked up (350 km) which includes the campground run;

2.) The Campground Run - the haul from Whistler and Wapiti campground as well as Jasper Tramway and the Youth Hostel;

3.) The East Run - the haul from Pocahontas, Miette Hotsprings, wayside bins and campground, the Snaring campground road and Maligne Lake (34O km).

See map of outlying collection routes (Figure 23).

All MSW hauled from the town travels the same distance to the transfer station, whether it is commercial waste or residential waste, collection routes. Collection times vary with the type of garbage being collected

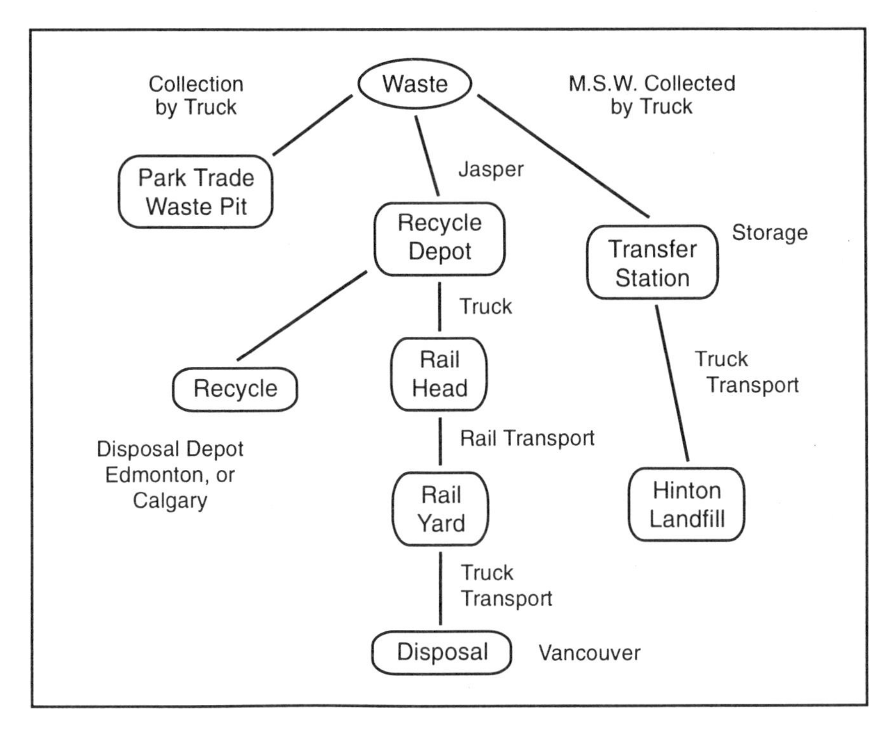

Figure 22 Disposal Location of Waste From Jasper National Park

and the number of non-hydraulic bins that must be emptied, rather than with the distance travelled. The principal garbage produced in town is either commercial or residential. Small disposal bins on the sidewalks were classified as commercial, as they are located only in the commercial section of town.

Recycled Goods

Materials suitable for recycling are found in both trade waste and MSW. Jasper has no internal recycling programs, therefore, the majority of this material must be either trucked or rail hauled out to a larger centre that has recycling markets. Recycled goods in Jasper have chiefly been glass, plastic, aluminum, tin and paper. Agencies responsible for ensuring that this material gets to a market are the park, the bottle depot and the senior citizens through their paper recycling program. There are no collection costs to any of these groups, as the material is brought in on a volunteer basis. Most of this material is taken to large recycling facilities in an urban centre.

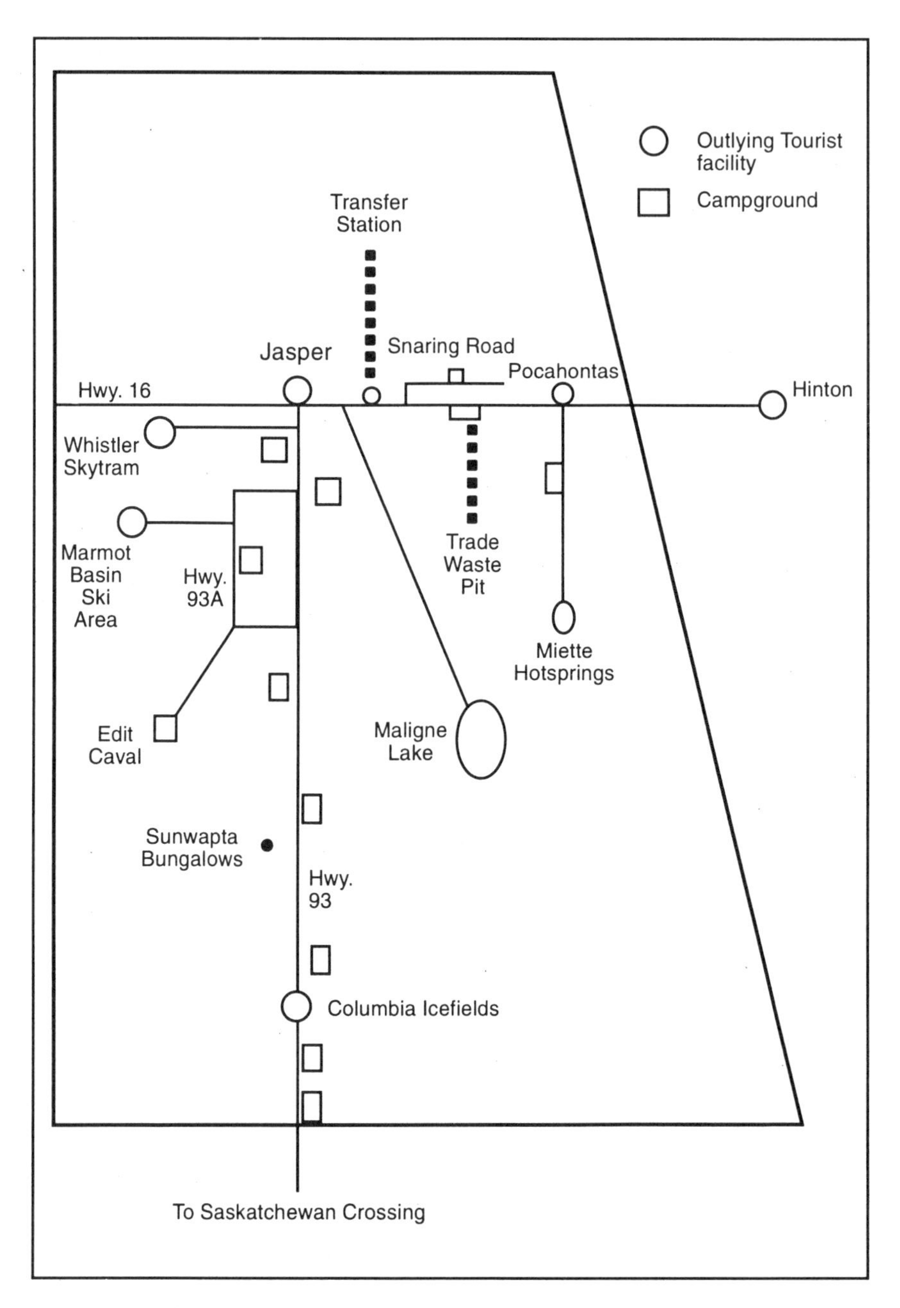

Figure 23 Collection Routes for Outlying Areas Jasper National Park

THE COST OF WASTE DISPOSAL IN JASPER NATIONAL PARK

The majority of the cost of maintaining a solid waste disposal system is taken up through collection, storage and transportation. The remaining cost comes from disposal (Messner Assoc., 1974). As cited earlier, collection costs alone may account for up to 80% of total costs (Cardile and Verhoff, 1974).

The government collects, transports and disposes of all waste generated in the parks, with the exception of recycled goods taken to the bottle depot and trade waste brought to the trade waste pit by the public. Since there has been no documentation, either by the government or the public of the amount of trade waste taken to the pit, it is not possible to calculate the cost per tonne to collect and transport this material to this location. However the cost can be calculated, to some extent for recycled goods. The cost to the government, for these two catagories, can be calculated through records kept on each truck used in hauling municipal solid waste and the hauling of recycled goods. The cost of hauling recycled material by private contractors is available through the individual companies operating in Jasper. At present the only waste handled by private companies is recycled material from the bottle depot.

Disposal Costs in the Park

Trade Waste

Although the Government keeps good documentation on how much it costs to run a specific vehicle, there has been no data kept on what or how much is hauled to the trade waste pit (pers. comm., Blaskco, 1992). Each truck has records of distance traveled, but not all of this mileage can be attributed to waste hauling. Weights are not recorded at the trade waste pit, making it difficult to assess the disposal costs. The pit operator does not work there full time and could only estimate the time spent on site covering new waste. The work load varies throughout the year, depending on the amount of material deposited. Given this variablility and the fact that no weights or records have been kept, it is not possible to determine the cost per tonne to dispose of this material at the trade waste pit. The best estimate for operating a dump truck comes from a study conducted in Alberta in 1982 on cost effectiveness of transfer stations versus direct haul (Bissell and Assoc., 1982). This study cites a cost of 19 cents per ton mile for operating an open dump truck with a crew of one and nominal loading of 1.3 tons (Bissell and Assoc., 1982). It must be noted that this study is ten years out of date and many of these figures will have changed. The data reported from this study kept the ton per mile context to maintain the integrity of the original report. An attempt was made to contact this company for updated information; however, it no longer exists.

The immediate cost of operating the dump can be arrived at by determining the time spent handling waste at the dump, the wages of the operator and the cost per hour of running the crawler dozer. The operator spends an average of one day a week disposing of the waste. In 1992, the operating cost of the crawler dozer was $8.49 per hour and the operator was paid $12.50 per hour. The total cost of operating the landfill crawler dozer per week was $209.90 This equates to an hourly operating cost of $21.00 while operating. The annual cost is approximately $10,915.00 (calculated from 1992 data). Note that the dozer is left at the site and therefore incurs no cost in moving from the trade waste

Without considering how much is hauled to the trade waste pit and using estimated operating costs determined by the Alberta Transport and the Hinton Landfill, transport of trade waste is estimated as 19 cents per ton per mile.

Although there is little information on the number of trips private contractors made to the pit, the Canadian Parks Service kept a record of the number of hauls the government made to the pit during the 1991–92

Table 29 Estimated Trade Waste Costs

Collection	Transportation	Disposal
Gvt rate loading cost	cent/ton/mile	$21.00/hr.
73 cents/tonne	19 cents/tonne	
(AB Transport)	(Bissel, 1982)	

fiscal year. Because the trade waste pit is rapidly filling up, Parks tries to deposit as much clean fill material in other locations as it possibly can. An example would be storing old fill that used to go to the trade waste pit at the government compound for other use in the future. The hauls that were taken to the pit are presented in terms of the season to remain uniform with other data on MSW (Table 30).

Parks service records kept on the operating cost of a tandem truck show that the typical tandem (truck #176269) in the park is run at 24.4 cents plus 8 cents depreciation (32.4¢) per kilometre. A single axle (truck #185519) costs the government 22 cents plus 8 cents depreciation (30¢) per kilometre. Unfortunately there is no record of where these loads were hauled from or the type of material hauled. Since weights are not available, it is not possible to calculate the tonnage, regardless of what material is hauled.

Calculating out the costs encountered in waste hauling for the private sector presents the same problem as with the government. None of the principal haulers have kept records on specific weights or numbers of hauls to the dump versus other hauling. However, the private contractor uses rates some what cheaper than that of Alberta Transport for the long haul (pers. comm., Oberg, 1993). Bowen Trucking is competitive with the other six companies in Jasper, charging 73 cents per tonne to load and 11 cents per tonne per kilometre to haul. There are no tipping fees at the trade waste pit, so disposal is not a cost (per. comm. Height, 1992). Note, the private contractor must achieve a payload of 17 tonnes per kilometre to remain economically viable.

Disposal Costs Outside the Park

Cost of Disposal of MSW

MSW differs in the handling costs from trade waste in that there is the added storage phase, entailing more handling of the garbage. This waste is weighed, allowing for costs per tonne to be calculated from the number of trips required to haul a known amount of waste to Hinton.

Evaluating Vehicle Costs

Evaluating the cost of operating a vehicle over a period of one week or one month does not give a true value because the depreciation and replacement cost of the vehicle must be factored in. In this study, using government documentation, the cost of operating a typical collection vehicle is calculated over the current life of that vehicle.

Parks Canada records the cost of all replacement parts, body repair, preventative maintenance, oil and fuel costs on a yearly basis. At the end of three months the cost of operating that vehicle is assessed and added to the total cost of operating that vehicle since its purchase. The final mainstem report includes a summation of normal maintenance and repair, downtime, road calls, money recovered from the warranty, any major modification, accidents, fuel and oil utilization and damage costs. These figures are summed up to give the total maintenance and repair and operating costs in dollars per every hundred miles of use. Because the government does all its own servicing, there are no additional costs in profit as encountered with private repair shops. Also the government carries no insurance on its vehicles and therefore insurance is not included in operating costs. A typical garbage collection truck would be a 1985 GMC truck with an allowed haul capacity of 13,500 lbs. (pers. comm., Blasko, G., 1992). Some of the figures on this truck are given in Table 31.

Table 31 shows that it costs the government 22.9 cents per km (the gov't reports this data in miles, which have been converted to km) to operate the above vehicle. This particular vehicle had a life of 6½ years before retirement. Depreciation costs are not recorded, but, may be calculated from the original cost of the vehicle, the number of years of operation and the final selling price. For the above vehicle the depreciation over this time period was:

> Original retail price minus resale value, divided by total number of kilometres.
>
> $15,512 - $2,500 = $13,012 / 159,273 km = 8.2 cents / km

Over this period the vehicle had been driven 159,273 km, therefore, the cost per kilometre for depreciation was 8.2 cents. The daily operating cost of that vehicle is 22.9 cents per kilometre. Adding the depreciation of 8.2 cents gives a total operating expense of 31.1 cents or roughly 3.1 cents per kilometre. The cost per kilometre has been calculated for the other vehicles used in handling or transporting MSW in JNP.

Collection Costs

Collection encompasses all those steps required to move materials from the source of generation to a transfer facility or a disposal operation (*National Waste Reduction Handbook*, 1991). This cost is difficult to establish due to the complexities involved. These variables include labour costs, the nature of the refuse, the collection method employed, road conditions, service density and the capacity of the trucks in service (Cardile and Verhoff, 1974). Collection costs for municipal solid waste differ from that of trade waste in that the vehicles used for waste pick up are not used for any other purpose. The actual cost of collection has not been calculated for vehicles used by the federal government, but

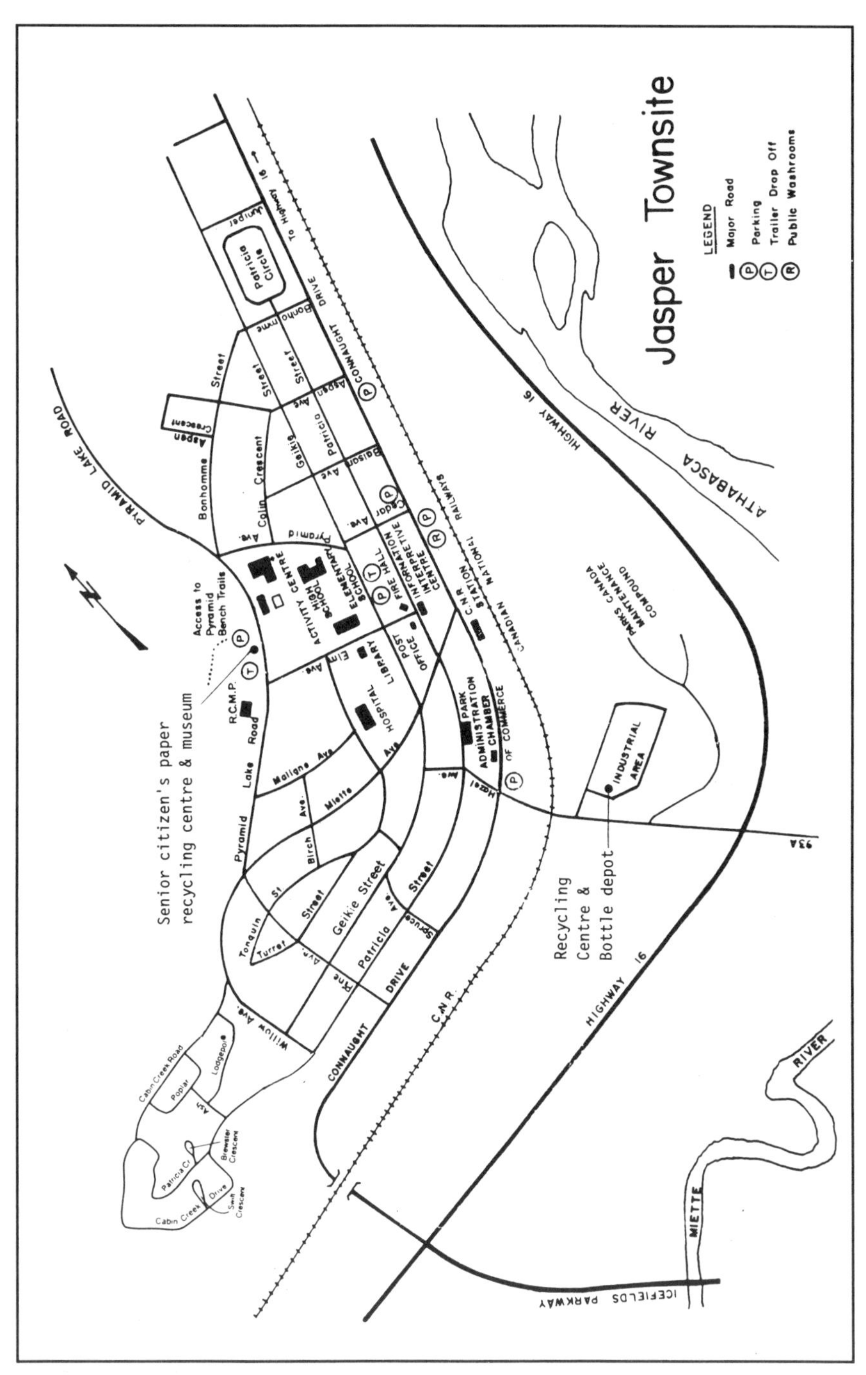

Figure 24 Jasper Townsite

may be estimated by considering the average wage of each driver and the cost of vehicle usage, per hour, while collecting MSW. These costs have been calculated at an average wage of $12.50 per hour and 31 cents per kilometre to run. Therefore, a driver employed for eight hours will use all of his time in either the pickup of waste or other related duties such as truck maintenance. From studies conducted on the waste stream of Jasper National Park at the transfer station, it is possible to determine the average number of loads taken from the townsite and outlying areas. The pickup on all routes is either from hydraulic bear proof bins or hand pickup from locally owned collection retainers. The time taken to empty garbage from both sources is estimated to be between 5 to 6 minutes per stop. As discussed in collection of trade waste, many factors affect this time, such as frozen bins, truck breakdowns, driver motivation, the weather and so on (pers. comm., Edwards, 1992–93). The amount of garbage to be picked up also varies with the season. The seasonal variation in pickup is illustrated in Figure 25. The park has a mandate to empty all commercial bins seven days a week and all residential bins six days a week. Should some bins not be full, they may not be emptied; however, a stop must be made at each bin to determine whether or not it needs to be emptied or not (pers. comm., Pigeon, 1993). The collection routine consists of driving from bin to bin until all bins have been emptied or checked.

Since it is difficult to separate the emptying process from the travel time between bin pick-ups without actually monitoring each run, the most accurate method to determine cost is to use the cost of operating the vehicle per kilometre and the salary dollars paid out during that collection period. The time spent during the day not actually engaged in pickup is minimal and, therefore, the assumption is made that the day is spent in the pickup activity. The number of runs was determined and the amount of waste collected was measured over a twelve month period between June 1, 1991 and June 2, 1992. Since it was believed that there would be differences in cost between each season, the calculation of cost per tonne to collect the waste was done for each season, according to those periods specified in the waste analysis study. This was determined by assessing the operational cost of running the vehicle, wages, mileage calculated from the number of runs made and the total amount in tonnes of garbage hauled. This cannot be done without making some assumptions:

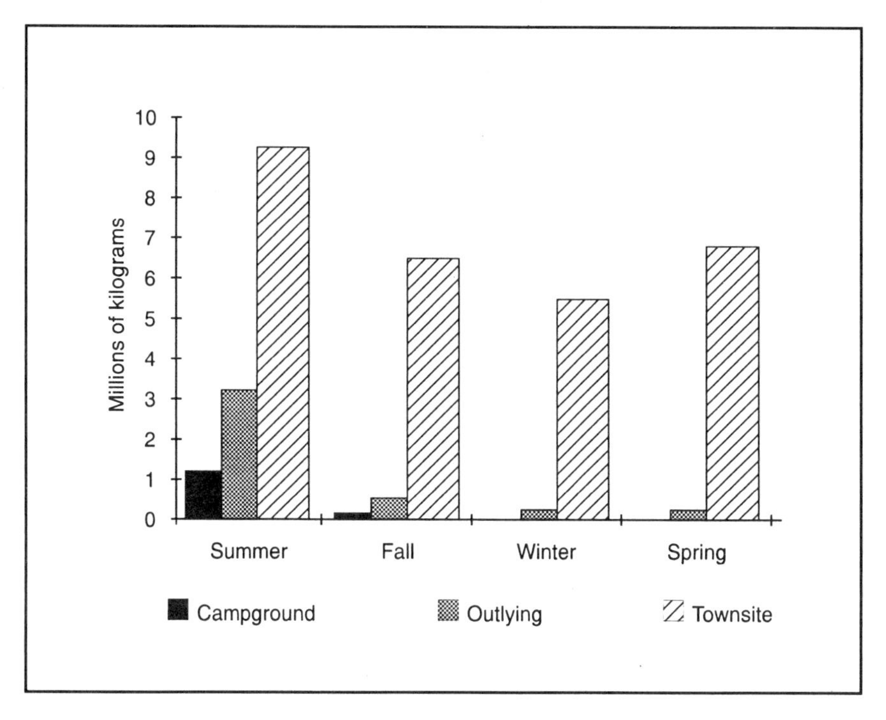

Figure 25 Total Weight Trucked By Seasons

1) The trip from town to the transfer station takes an average of 45 minutes in the summer to one hour and 20 minutes in the winter. At $12.50 per hour, the summer trip costs $9.38 in wages. The winter trip costs $16.67 in wages.
2) The average number of kilometres driven for each town trip is 24.
3) The average out of town round trip takes eight hours, giving an average driver cost of $100.00 per trip.
4) The average number of kilometre travelled on out of town trips is 350.

The vehicle cost is arrived at by taking a typical collection vehicle's mainstem report and converting the cost per 100 miles to kilometres, then breaking it down to a per kilometre cost. The depreciation cost, calculated to be eight cents per kilometre was added to the operation cost.

The cost was calculated only for summer and winter, as these seasons are at the opposite ends of the cost spectrum. The difference was calculated for both in town and out of town transportation of waste. The results are as follows:

Summer in Town = $0.80 per tonne per kilometre
Winter in Town = $1.34 per tonne per kilometre
Summer out of Town = $0.69 per tonne per kilometre
Winter out of Town = $0.76 per tonne per kilometre

The number of trips taken and the amount hauled is taken from Table 33.

Table 30 Government Loads Taken to Trade Waste Pit

1991 - 1992		
Summer	(June, July, August)	252 loads
Fall	(Sept., Oct., Nov.)	245 loads
Winter	(Dec., Jan., Feb.)	110 loads
Spring	(Mar., Apr., May)	34 loads

Summer (town): From Table 32 the number of trips taken was 301 and the amount hauled was 263,959 kg or 264 tonnes.

Winter (town): From Table 32, the number of trips taken was 247 and the amount hauled was 153,366 kg or 153 tonnes.

Summer (out of town): From Table 32, the number of trips taken was 102 and the amount hauled was 87,687 kg or 88 tonnes.

Winter (out of town): From Table 32, the number of trips taken was 5 and the amount hauled was 3,900 kg or 4 tonnes.

The main observation for the park is the disparity in the amount of weight hauled in the summer per number of runs versus the amount hauled in the winter per number of runs. Clearly, hauling MSW in the winter is not cost efficient, particularly for town pick up.

The time taken for pickup (45 minutes per trip in the summer, one hour 20 minutes in the winter) can be accounted for by the conditions under which the pickup is made. Winter conditions are, on average, more difficult for travel.

Handling Costs at the Transfer Station

The park has one person working at the transfer station ten hours a day, seven days a week, all year round. The work consists mainly of unloading the collection vehicle and reloading the waste into a waiting transfer trailer. This person will drive the refuse to Hinton when the trailer is loaded. The annual cost of running this station can be calculated only over a two year period, as that is the length of time it has been operational. This cost does not vary from month to month. What does vary is the amount of garbage handled for the same costs per month. As with trade waste, MSW is seasonal and varies greatly from summer to winter. The cost of handling is best expressed as the average cost of handling one tonne of garbage for each season. The cost of running the station per season is found by dividing the annual cost by four. This cost, divided by the total number of tonnes shipped to Hinton in that time frame, yields the cost per tonne, per season.

These figures come from the 1992 operational cost analysis of the transfer station done by the park (pers. comm., Blanchett, 1993). The number of tonnes processed through this facility is available through the Highway and Scenic Corridors section of Parks Canada, from the computer recorded weights taken from the transfer trailer just before each shipment to Hinton. The amounts are recorded in kilometres and the monthly total is recorded for the year. The rates are given in Table 31.

Given that there are four seasons, the cost to operate the transfer station is therefore $10,449 per season. The amount of waste stored and handled during each season varies considerably from summer to winter.

Transportation to Hinton

The cost of transporting MSW to Hinton includes the cost of operating the truck and the driver's salary. By calculating the amount hauled in one month and dividing by the cost per trip, a cost per tonne can be arrived at.

Each trip takes three hours and the trailer travels 72 km one way, or 144 kilometres return. The driver is paid $12.50 per hour (including compensation and UIC). The cost of operating the vehicle is 50.8 cents per kilometre (this is the cost of the tractor at 33 cents per kilometre and the tranfer trailer at 17 cents per kilometre). The depreciation has not been calculated on this vehicle, as it has not been sold, but, by assuming it should be similar to the depreciation on the garbage truck at eight cents per kilometre. As with the transfer station, this is considered on a seasonal basis. Refer to Table 32 for total cost per trip.

There is less disparity from season to season in transporting waste to Hinton than in the handling or collection costs, mainly because the trailer is usually loaded to capacity before a haul is undertaken. This changed, however, in the spring when more frequent trips with an underloaded

Table 31 Truck Maintenance Costs

Maintenance and Repair		$23,003
Down time (cost of not operating vehicle)		$606
Road calls (number of rescues)		3
Normal wear (parts and labour)		$22,228
Warranty		$26
Damage (damage costs to vehicle beyond normal wear)		$749
Accident		$520
Capitalized statutory (major modification, new chassis)		$3,826
Recovery		
Fuel		$13,106
Add oil (new oil added)		$129
Fuel gal (cost of fuel per mile)		10.7¢
Per qt (# of miles obtained on one qt oil)		1,207
Cost per 100 miles		
Maintenance and repair	$23.24	
Operating	$13.37	
Total	$36.61	
(1992 data, per unit)		

trailer were taken. At this time of year, due to the frost coming out of the ground, there are road bans on allowable weight to be transferred and the park is restricted to 70% of their normal load. This will change in 1994, when the road will be upgraded to allow maximum transport. Until then, the cost per tonne will be as calculated.

Disposal Costs

The disposal costs at the Hinton landfill take into consideration the unloading time, disposal, wages and machinery expenses. From these daily operational costs, a tipping fee is calculated. All dumps have a standard fee, but not all municipalities take into consideration how fast the dump area will fill or how expensive it would be to relocate. The original costs for operating the landfill have been re-estimated and the tipping fees have risen. Currently, Jasper pays $16.00 a tonne but this will go up to $19.50 a tonne in April, 1993 (pers. comm., Polski, 1993). The costs of

Table 32 Government Vehicle Operation Costs

Vehicle	Function	Cost/km
garbage collection vehicle (#195452)	remove MSW to transfer station	31 cent/km.
front end loader (#240005)	used to load transfer trailer at transfer station	$4/hr ($2.48/km)
transfer trailer (#232804)	removal of waste to Hinton	17.8 cents per km.
tractor (#195924)	the hauling unit of the transfer trailer	33.0 cents per km.

* Note the original values were given in miles. The above figures are converted to kilometre. The tractor and trailer expenses were retrieved from the mainstem reports on the individual units.

running the landfill do not change from day to day, regardless of the amount of material brought. The question of cost per tonne, however, must still be considered, because the amount of Jasper's garbage changes dramatically from summer to winter. Again, there is the problem of efficiency, which may be viewed as the cost to handle or dispose of one tonne of garbage throughout the year. Investigation into these costs is not part of this study, and the tipping fee will be considered the cost of disposing of one tonne of garbage.

Cost of Private Disposal of Recycled Goods

Collection

Recycled goods differ from collection of MSW in that there is no structured collection process in Jasper, outside of providing a recycling centre that goods can be brought to by the public. Collection costs for this material cannot be determined by a structured system, where cost per tonne per kilometre, can be established. Most rural communities in Alberta have a blue box curbside recycling program or a central collection system in place, and the recycled material is then picked up by the municipality (The National Waste Reduction Handbook, 1991). This program has not been established in Jasper; therefore, the costs of bringing in recycled goods to the collection depot is a cost to the public. The three locations

Table 33 Seasonal Variation in Kilograms Per Number of Loads Hauled

Season		Town		Out of Town		Total
	kgr	Loads	kgr	Loads	kgr	Loads
Summer						
June	78,014	94	11,932	18	89,946	112
July	82,445	109	37,405	41	119,850	150
August	103,500	98	38,350	43	141,850	141
Total	263,959	301	87,687	102	351,646	403
Fall						
Sept	71,892	88	11,154	15	83,046	103
Oct	66,589	75	3,531	3	70,120	78
Nov	56,391	68	3,020	4	59,411	72
Total	194,872	231	17,705	22	212,577	253
Winter						
Dec	49,800	84	2,020	3	51,820	87
Jan	51,622	80	1,000	1	52,622	81
Feb	52,240	83	900	1	53,140	84
Total	153,662	247	3,920	5	157,582	252
Spring						
March	53,320	78	1,350	3	54,670	81
April	58,250	81	1,920	3	60,170	84
May	85,083	90	6,411	8	91,494	98
Total	196,653	249	9,681	14	206,334	263

are the bottles depot, the recycling bins in the Stanley Wright Industrial Park and the Senior Citizens Paper Recycling centre near Jasper's Yellowhead Museum (see map of Jasper townsite, Figure 24).

Cost of Disposal – Bottle Depot

Storage costs

The bottle depot takes in glass, aluminum cans and plastic disposable bottles. With the exception of Alberta Brewery beer bottles, hauled by ABA, all the recyclable containers are hauled by Grimshaw Trucking in Hinton to the Containaway disposal plant in Edmonton. This material is sent on average once a week, so the handling and storage costs were figured out for weekly periods. The depot kept records of all the material sent out over the first year of operation. From this it was possible to determine total weights for each season, and for the whole year. The owner also had costs on salary dollars per year, rent per year and utility costs per year. The total weight was calculated per week and divided by the operating expenses per week, to give the cost to handle and store one tonne of material per week.

In one week the bottle depot handles and stores an average of 14.3 tonnes of material. The depot manager estimates that it costs him $950.00 to run the depot over a period of one week. This gives an average operating expense of $66.0 to handle one tonne of material in this time period.

The material is shipped to Containaway, where there is more handling and storage before it goes on to a final disposal site. Containaway is a private company and would not provide information on any costs they incurred in the process of storage, handling and shipping. With some reluctance, destinations of the following materials were given (Containaway, 1993).

Aluminum: Sent to the United States; specific location not given.
Glass: Sent to Canisfer in Calgary for final reprocessing.
Plastic: Remains in Edmonton for recycling.

Transportation Costs

The costs of operating a private vehicle are not recorded in the same detail as the government records, however, in order to estimate the cost of running a trucking business, the following costs must be considered:

Truck size: Hauling capacity
Maintenance: Parts and labour
Shop rent: Cost of housing the vehicle and land rental
Wages: Salary of persons employed plus workmans compensation and UIC
Fuel and Oil: Costs incurred to run the vehicle
Depreciation: Monitary loss through breakdowns
Insurance: Cost to ensure the vehicle
Vehicle replacement: Cost of new vehicle

For a typical trucking company in Jasper such as Bowen Transport, these costs can be considerable. This company operates one 15.7 m triple axle trailer, a 14.8 m trailer and one forklift. The company employs three operators, who including workmans compensation, costs them $8,400.00 per month. The insurance costs are $4,000 per year, the fuel and oil costs were $15,800 for the year and normal repairs ran $27,500 for the past year, 1992. Vehicle replacement cost averages around $100,000.00 for a new truck. The depreciation was 30% for both trucks in 1991–92. The purchase cost of the trucks was $33,439 and $22,250, respectively, giving a depreciation cost of $10,037 and $6,675 in the first year (pers. comm., Height, 1993). Although depreciation may be taken off income taxes, this cost is still a loss and is compensated for only by receivable money through tax rebates. Thus the total yearly operational costs would be $145,300 for the year, minus tax write-offs.

All recycled material processed through the bottle depot is shipped by Grimshaw Trucking out of Hinton. The company operates two 14.8 m trailers, one of which is on site at all times for the purpose of loading and storage. Part of the handling costs are actually loading costs, which the company does not get paid for. The company was very reluctant to disclose cents per tonne per km costs, but did say that each run to Edmonton cost approximately $400. The distance to Edmonton is 362 km (one way), which would yield a cost of $1.01 per km. This figure was given as an average cost, but, does vary from load to load depending on whether the material is largely plastic, aluminum or glass. The company is reimbursed from Containaway once delivery is made. The profit and therefore the cost of hauling depends on the price of the material hauled. At present plastic bottles are worth more than glass, which is greater in weight, but less profitable to haul (pers. comm., Goche, 1993). No information was forthcoming specifically on transportation of materials once it left the Containaway facility. Most trucking companies when bidding on longer hauls, usually bid about 80% of the standard government rate. Many of the larger recycling companies such as Containaway and Paperboard use several trucking companies to haul all the material received.

Senior Citizens Paper Recycling

Storage Costs

Jasper residents bring paper to the storage trailer on their own volition, leaving no collection cost to the senior citizens. The trailer that houses the stored paper was purchased from Allied Paper in Edmonton, through a grant from the federal government. After being used to haul paper for three years, the trailer will be donated to this volunteer group. All work is done by volunteers, therefore, handling and storage is cost free.

Transportation Costs

The transportation of the paper was previously handled by a private trucking firm which charged $400 a load. In the last year, the government took over this job without a charge, so this cost reverts to the government. Prices quoted by Dave Edwards, (1993) was again $364 per trip. The paper is hauled to Edmonton (362 Km). Each load averages out to 20 tonnes, giving a cost of 2.5¢ per tonne per kilometre.

Cost of Government Disposal of Recycled Goods

Storage

Jasper Recycling Society has provided 8 bins at the Stanley Wright industrial park, where people can bring glass, tin and aluminum. These bins cost the Society $3,000 for each bin. This is an average of $6,000 per year over the past four years. Very little material is recycled through these bins as it is more profitable to take these goods to the bottle depot. Pick up time is short and the bins are emptied approximately once a month. This material is taken to the transfer station where it is stored in concrete bins, until enough has accumulated to warrant a trip to the recycling center. The principal costs are those of the bins and the handling costs at the transfer station. The operator at the transfer station crushes the tin when he is otherwise unoccupied loading the transfer trailer. This time is erratic and not documented.

Transportation Costs

There are two transportation costs:

1) hauling the recycled material from the town bins to the transfer station;

2) transporting the material from the transfer station to the disposal site.

The cost of hauling the recycled material from the town to the transfer station is difficult to determine on a tonnage basis, as these trucks were not weighed when they brought in a load. The pick-up is determined by how frequently the bins fill up. The cost of operating the collection vehicle, however, would be the same as the cost per km of operating the

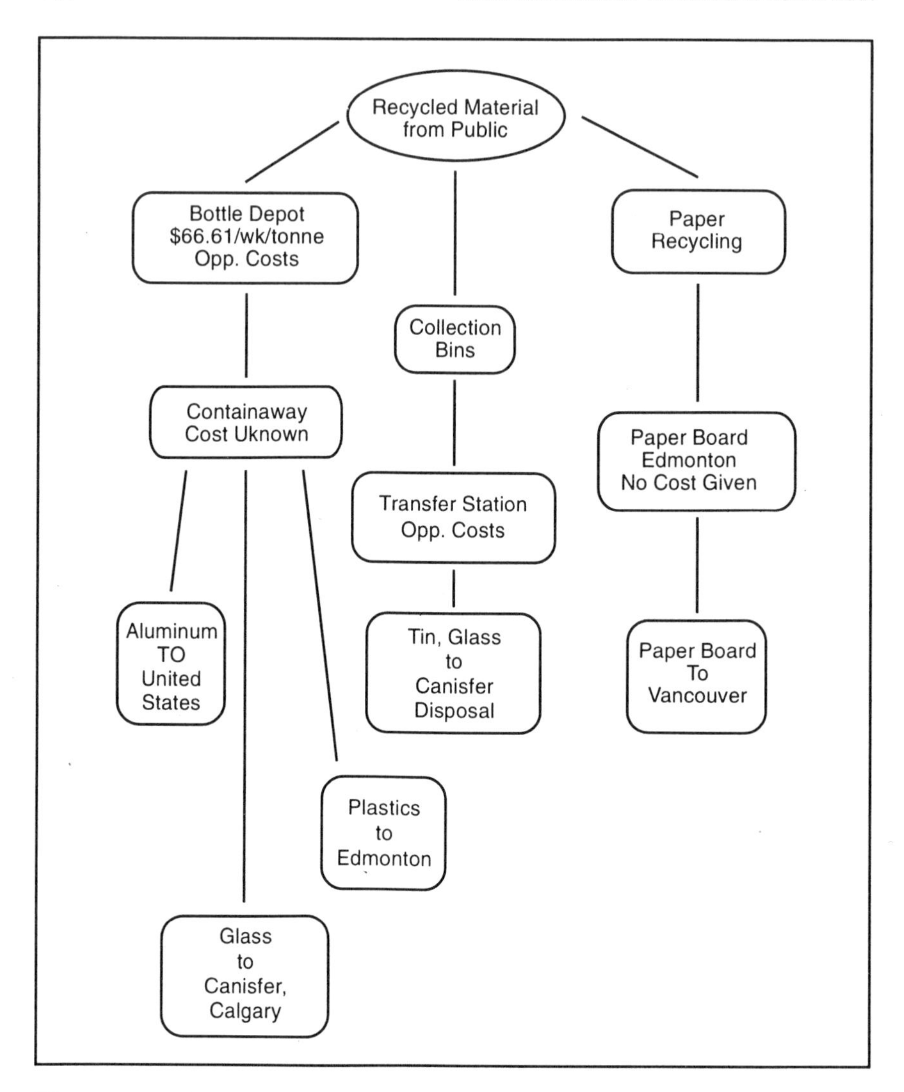

Figure 26 Disposal Destination of Recycled Goods

MSW collection vehicle. Parks Canada hauls glass recyclables to Canisfer in Calgary. With one exception, tin has been transported to Stelco Industries in Edmonton.

It is possible to estimate the travel costs to the recycling centres as the loads are weighed before they leave the transfer station. There is a discrepancy in the cost of handling paper, versus glass or tin, in that the paper weights hauled to Edmonton in one load are considerably greater than the hauled weights of tin or glass. The vehicle and operator costs are the same for both hauls. As a result, paper is much cheaper to recycle than tin or glass. The problem with hauling tin and glass is that they are both light materials and it is difficult to reach a maximum pay load.

A list of the number of loads hauled to Edmonton or Calgary and relative costs are given in Table 37.

Table 34 Cost of Operating Transfer Station

Wages, plus overtime	$33,988
Maintenance	$5,305
Operation of Backhoe (cost $4/hr)	$2,185
Total	$41,478

* Total wages are $48,554, 70% ($33,988) spent at transfer station and 30% ($14,566) spent driving.

Table 35 Waste Handling Cost of Transfer Station Per Season

	Summer	Fall	Winter	Spring
Operation costs	$10,370	$10,370	$10,370	$10,370
Tonnage handled	1376.6	763.8	548.7	720.7
Cost/tonne/season	$7.53	$13.58	$18.90	$14.39

Table 36 Seasonal Cost of Hauling Waste to Hinton

	Summer	Fall	Winter	Spring
Number of Trips	85	51	37	69
Number of km	144	144	144	144
Vehicle Cost per km	$0.51	$0.51	$0.51	$0.51
Cost per Trip	$73.44	$73.44	$73.44	$73.44
Total Cost	$6,242.00	$3,745.00	$2,717.00	$5,067.00
Driver's Wage per hour	$12.50	$12.50	$12.50	$12.50
Number of Hours per trip	3	3	3	3
Driver's Cost*	$3,187.50	$1,912.50	$1,387.50	$2,587.50
Total Operating Cost	$9,429.50	$5,657.50	$4,101.50	$7,654.50
Tonnes Hauled	1376.7	763.8	548.7	720.7
Cost per tonne	$6.85	$7.41	$7.48	$10.62
Cost/Tonne/km.	4.2¢	4.7¢	5.1¢	10.2¢
Avg. Wt. hauled per trip	16,196 kg	14,976 kg	14,828 kg	10,444 kg.

Avg. Cost per Tonne = $7.87
Avg. Cost a Tonne per km = 5.5

* Driver's cost = No. of trips × hourly wage × hours per trip

Table 37 Government Cost to Transport Recycled Goods

Date	Tin	Aluminum	Clear Glass	Green Glass	Brown Glass
Mar 13/90	1182 kg		1558 kg		
Jul 24/90	1591 kg				
Feb 27/91	2318 kg		2831 kg		
Mar 19/91					496 kg
Jul 24/91			1448 kg		
Dec 10/91	2555 kg				
Apr 07/92			7814 kg		
June 92	2964 kg				
Oct 92	3464 kg				
Total	14,074 kg		13,651 kg		496 kg

Calculations on transportation of recycled material (Government):

Number of Loads Hauled & Milage	Material Hauled	Cost to Run Vehicle/Km.	Driver Cost/Hr	Amount Hauled	Cents/ Tonne/Km.
Round trip Cal. 832 km 3,328 km	Glass	35.4 cents/km	$12/hr 12 hour-trip 4 trips	14,147 kg	14.9 cents
Round trip Edm. 724 km 4,344 km	Tin	35.4 cents/km	$12/hr 10 hour-trip 6 trips	14,074 kg	22.2 cents

7 Recommendations

Many recommendations are implicit in this study's conclusions; however, these are broad statements that do not indicate specific actions. To identify the problems inherent in proper waste management, it is essential to develop a waste management plan. The following recommendations should be addressed in this plan, but, initially, it would be expedient to write an interim management plan to deal with immediate problems.

THE INTERIM MANAGEMENT PLAN

The most pressing problem in the park today is the management of trade waste. The present trade waste pit will be closed in the near future, as there is no more room at this site. Also, it is in a poor location established prior to the development of guidelines set up to prevent contamination associated with landfills. Specific recommendations for closure and re-establishment are:

- Identify requirements for closure by following the guidelines established by the Province of Alberta.
- Identify a new site, also following the Alberta guidelines.
- The new trade waste site should be enclosed and monitored. This would require a gate, fence and a staffed position.
- Persons wishing to dispose of trade waste should require a permit. The cost of the permit should reflect the composition and amount of the material to be disposed of.
- Scavenging now occurs at the present trade waste pit. Because of open hours and lack of security, many people go to the dump to recover salvageable material. Although, this is currently against the Parks Canada garbage regulations (National Parks Act, 1980), scavenging is a form of recycling and thus is a positive activity that should be endorsed by the park. If a permit system were

established this could continue. This recommendation proposes that persons wishing to salvage material be issued a permit to do so and be given hours in which this can be accomplished.

Much of the trade waste is dirty material from old fillsites. The Government is diverting this material as effectively as it can to other locations and uses, but, it would be expedient to establish a filtration system to clean and crush the old gravel received from government or private sources to replenish gravel supplies rather than allowing it to become trade waste.

Currently there are recycling areas established at the Sanitary landfill for such goods as tires, wood and appliances. This, however, is not on the trade waste site and requires an extra diversion for persons wishing to recycle these goods. The new trade waste pit should establish areas to accomodate these goods and ensure that the public is well aware of the facility.

The interim guidelines should also indicate ways to reduce the quantity of compostable material brought into the trade waste pit. It is common for some landfills to refuse yard waste, requiring the individual to dispose of it elsewhere. Yard waste is not as significant a factor in Jasper as in some parts of the United States and has not been excluded from the trade waste pit.

MUNICIPAL SOLID WASTE

The interim management plan must contain guidelines for dealing with reduction of MSW. This includes recycling, as increased recycling will reduce the amount of material entering the waste stream. The major sources of garbage in the park comes from the commercial section. Of this a high percentage is food and cardboard waste. Residential waste is the second highest contributor, but the waste stream differs significantly in composition from commercial waste. Campground waste is similar to residential waste. The following recommendations are directed toward reducing MSW.

All commercial facilities that deal with food must separate materials that contaminate a food source that may be composted. It is recommended that small plastic containers used for cream, butter or jams be replaced with jugs or jars in their place. This would eliminate tedious separation of this material before waste is dumped in the waste bins.

Commercial facilities should attempt to buy in bulk. This reduces the number of containers used and is cheaper. However, bulk material can be hard to come by. It may be possible for all food outlets to combine their resources and order in bulk from one commercial supplier who deals in bulk commodities.

All commercial outlets should investigate the amount of packaging their suppliers use. Ordering supplies using a minimum of packaging would not only reduce the cardboard content, but doing so would also greatly reduce the amount of plastic in the waste stream. It is recommended that a small committee be put together to research suppliers of minimum packaged products. This information can then be passed on to the business community.

An audit should be made of the amount of money the Park spends on handling and disposing of waste over one year. This can be compared with the total revenue received from the community for the service. The interim report should recommend this as a separate study from the report. The study should focus on the benefits of increased cooperation with the community in reducing their waste versus an increase in tax revenue to compensate for the costs of waste disposal.

An education program should be established to make the public aware of how much waste and its composition is being disposed of each year. This program must be ongoing and include specific solutions to specific problems.

Composting is a must. To obtain a sustainable compost program it will be necessary to do further research in this area. It is recommended that the interim management plan address the specifics of starting a compost program in the park. The full planning of a compost program, however, should be outlined in a long term waste management plan.

Transportation of waste has been identified as costly to the park. Some initial cost figures have been given in the chapter on Transportation, but this information should be refined. At this point it is recommended that Jasper National Park continue to be responsible for transportation of this material, but with a defined goal to increase efficiency. This may result in a modified collection routine and modified work hours. A secondary study on this issue is recommended.

The composition of residential waste will not likely change; however, the quantity can be reduced by increased recycling and discerning purchase practices. The resident, who is a permanent customer, can influence product-packaging practices. It is recommended that a purchasing guide be written to indicate what products are over packaged and what the alternatives are. Significant public pressure will change what is available on the market.

Yard waste is a prime seasonal component of residential waste. At this time residents have no alternative to dumping accumulated yard waste. It is recommended that specific collection depots be established to receive this material, once a composting system is in operation. The education program must emphasize that yard waste not be contaminated with glass, tin, aluminium, plastic or any hazardous materials.

The compost program should indicate how residential food waste may be handled by the system. A long term plan for this should be identified in a food waste management program.

Organic other waste is high in residential garbage. Other household material, such as used furniture, goes to the trade waste pit. A flea market could be established to recycle these goods through the community. Jasper sponsors an Environmental Week, each year, in the spring. It is recommended that a week long flea market be established to recycle as much of this material as possible during this period.

At present Parks Canada's recycling bins are at a separate location from the bottle depot. It is recommended these bins be moved to the depot location to facilitate recycling and provide a more convenient location for the public.

A waste management plan should be prepared addressing long term goals for waste reduction. This will require projecting costs and population dynamics. Jasper National Park currently forecasts projects and developments over a five year period. These projections could be used in developing a management plan that will project the increase or decrease in specific wastes in the next few years. A sustainable waste reduction program would be the goal of this plan, this plan, must take into consideration the seasonal nature of waste generation in the park. Elements in the program should be tailored to the fall, winter and spring seasons.

8 Conclusion

The purpose of this study was to determine the source and amount of all waste generated in Jasper National Park, in order to devise a waste reduction program to meet the goal of Canada's Green Plan for waste minimization. The conclusions presented here reflect the factor which influence the possibility of attaining this goal.

TRADE WASTE

Trade waste is the only waste disposed of within the park. Because this material cannot be taken to the Hinton Regional Landfill, disposal of trade waste will continue to be a problem for the park. The present trade waste pit is rapidly filling up, and a new location must be identified in the coming year (1994). This relocation will allow the government to rectify the problems associated with the old trade waste pit. The design of the new pit must be well planned to meet the demands of the type of material that will be disposed of there.

It will be important for the government to develop a plan that addresses such concerns as hours of operation, access, permit requirements, materials allowed, manpower, security and longevity of the site location. This plan should also address future rehabilitation potential. It will be important for the government to work with the community to ensure a smooth transition from random unsupervised dumping, now currently allowed, to a properly controlled facility.

MUNICIPAL SOLID WASTE

Because of the facilities and controlled access at the transfer station it was possible to do an in-depth study of MSW. The amount of waste handled, as recorded by the government, indicates that the park transfers around 4000 tonnes of MSW annually to the Hinton Regional Landfill. The only possibility of reducing this amount is to identify the major contributing elements and the sources. Examination of the Figure 20, and Tables 5 and 12, leads to the conclusion that a substantial reduction in compostable material (food and yard waste) and paper products (paper and cardboard) would result in a 50% reduction in total exported waste.

As a result of this study, the park has developed a program to recycle cardboard. This project was implemented without the development of an action plan for overall waste minimization. The next step should be the development of an interim waste management plan that establishes long term goals and short term objectives. Objectives should be established early in the planning process as a guide for a program design. This will require the participation of the community to identify waste management the assets and limitations of the whole park. The interim plan will begin initiation of programs that can be further developed in future. Goals should be identified in a comprehensive waste minimization plan that includes long term projections of anticipated changes in the waste stream. The trade waste program should be included in the final plan.

RECYCLED GOODS

Most of recycled goods, aside from cardboard, are handled by the community. This is material that has entered the waste stream and must be addressed in a waste minimization plan. For this reason the community should be involved with any program developed by the park. The major problem with recycled material is that of obtaining markets. Only materials that have secure markets should be considered for collection in a recycling program (Siedel, 1992).

One avenue of recycling waste is to compost food, ash and yard waste. Should this waste be converted to a usable compost, the potential waste reduction could be 25% of the waste stream. Without composting it is doubtful if a 50% reduction in waste by the year 2000 is possible. To accomplish this it will be vital to include the community in any program developed. Food waste is highly contaminated when it reaches the transfer station and must be source separated before it can be composted safely.

COST REDUCTION

Reducing costs can be critical to the development of a long term waste management plan and must be addressed in the plan. As indicated in Chapter 6, the single biggest cost to the park in disposing of waste is that of transportation. Because Jasper is an isolated community and has no internal recycling program, all recyclable material must be trucked out. A waste reduction plan should address ways to recycle material within the community where possible. Composting reduces this cost through lowered trucking and tipping fees. The compost would also replace the need to bring in topsoil from such sources as Hinton and Edson. This is also true for glass, which can be crushed and used as cullet for sewage lines. It is important to reduce transportation costs through efficient handling of material that is time consuming to load. This may mean increased equipment costs, but it will also result in reduced labour costs and will enhance the programs long term sustainability.

EDUCATION

The above conclusions emphasize the need to involve the community in any park endeavors to reduce waste. This can be most effectively accomplished by establishing an ongoing education campaign. Effective promotional activities will increase participation and reduce contamination in the recycled waste stream, while promoting overall environmentally responsible behavior. An education program should be included in any master waste management plan developed by the park.

THE RESPONSIBILITY OF THE PARK

Canada's national parks are set aside to preserve our wild areas for the protection of ecosystems and for the education and enjoyment for future generations. The government has dedicated these lands to the public by stating in the National Parks Act:

> *The National Parks are hereby dedicated to the people of Canada for their benefit, education and enjoyment, subject to this Act and the regulations, and the National Parks shall be maintained and made use of so as to leave them unimpaired for the enjoyment of future generations. 1930, as amended, 1985. c.33, s.4.*

The Government of Canada has publicly declared its responsibility for preserving these lands as close as possible to the condition in which they were found (*National Parks Policy on Reserve Protection,* 1983). The effect of accumulated waste is now being felt in the park, particularly since the park must support the burden of waste generated by a seasonal visitor population. The park has a responsibility to the public to be a leader in this field and set an example in preservation by disposing of its waste in a sustainable and environmentally sound manner.

Bibliography

Back to the Bay State, 1978, Solid Waste Management Refuse Removal Journal, v. 21(10), p. 22-26.

Barns, J., 1974, *Railhaul Transportation Role in Regional Recovery Disposal Systems - One Cities Solution*, Bureau or Polution Control, Atlanta, G.A., Compilation of Papers, San Francisco, California, p. 181–186.

Benjamin, J., and C, Cornell, 1970, *Probability , Statistics and Decision for Civil Engineers*, McGraw, Hill, N.Y.

Bird & Hale Ltd., 1976, *Development of a National System to Inventory General Municipal Refuse*, v. 2, Technical Report.

Britton, D.W., 1972, *Improving Manual Solid Waste Separation, ASCE,* 98(SAS), p.717-730.

Bissel and Assoc. , 1982, *Transfer Station Manual, Revised Addition,* Sponsored by Alberta Environment.

Burnett, M., 1991, *Composting Yard & Food Waste at the University of Calgary, An Action Plan*, M.D.P. for the Faculty of Environmental Design, University of Calgary.

Canada's Green Plan 'In Brief', A summary of Canada's Green Plan for a Healthy Environment, 1990, Cat. No. EN21-95-/ 1990E.

Cardile, R., Verhoff, F., 1974, *Economical Truck Size Determination*, American Society of Civil Engineers, Journal of the Environmental Engineering Division, v. 100, EE3, p. 679-697.

Code of Good Practice on Dump Closing or Conversion to Sanitary Landfill at Federal Establishments, 1977, An Environmental Protection Service document, Report No.: EPS 1 EC 27-4.

Curruth, D.E & A.J. Klee, 1969, *Analysis of Solid Waste Composition, Statistical Techniques to Determine Sample Size*, US Dept. Health, Education & Welfare, Public Health Services, Washington, DC, p. 25.

Devries, A. & B. Ross, 1990, *Analysis of the Policy Implication of Regional Municipal Solid Waste Composition.* Science Application International Corp., Sponsor: Ross and Assoc., Seattle, Washington.

Dillon & Assoc., Consulting Engineers and Planners, 1981, *Jasper, Edson Waste Management Study*, in association with Bissell & Assoc., for Alberta Environment, file no. 8820-01.

East Calgary Regional Concept Plan, Study of Recreation After-use at the East Calgary Sanitary Landfill, Prepared by Landplan Assoc. Ltd., Golder Assoc. Reid Crowther and Partners Ltd.

Edwards, F., 1981, *Sanitary Landfill; Legislation, Environmental Protection and Quality of Life Issues*, Alberta Environment, p. 18.

Ettala, M., 1991, *Revegetation of Industrial Waste Disposal Sites*, Waste Management & Research, v. 9, p. 47-53.

Ferguson,B., 1978, *Transfer of Solid Waste By Long Distance Rail Haul*, G.L.C., Public Health Engineering, Division Engineering Department, Solid Wastes, v. 68(9), p. 427-440.

Free, B., 1986, *A Social Perspective of Recycling in Alberta*, Environment Council of Alberta, Edmonton, Alberta.

Freeman, A.M., 1973, *Economics of Environmental Policy*, John Willey and Sons, Inc., New York, NY.

Gadd, B., 1986, *Handbook of the Canadian Rockies*, Corax Press, Jasper, Alberta, T0E 1E0.

Goldin, H.J., 1987, *"A Garbage Recycling Program for New York City"*, Journal of Environmental Systems, v. 17(1), p. 47-64.

Gore, A., 1992, *Earth in the Balance, Ecology and the Human Spirit*, Houghton Mifflin Co., 2 Park St., Boston, Massachusetts, 02108.

Guidelines for Industrial Landfills, 1987, Prepared by Environment Protection Services, Alberta Environment.

Hays, B., 1978, *"Repair, Reuse, Recycle—First Steps to a Sustainable Society"*, World Watch Paper, v. 23.

Hoi Tink, 1990, *Compost Symposium*, Olds College, Olds, Alberta, May, 1990.

Howlett, J., 1990, *Guidelines for Household Solid Waste Recycling Programs*, Department of Environmental Design, University of Calgary, Calgary, Alberta.

Hughes, B. 1992, *Railhaul Step by Step*, Waste Age, October Edition.

Klee, A., 1980, *Quantitive Descision Making*, v. 3, Ann Arbor, Scwin Publishers, Michigan, p. 40-57.

Klee, A. J. & D. Carruth, 1970, *Sample Weights in Solid Waste Composition Studies,* Journal of the Sanitary Engineering Division, Proceedings of the American Society of Civil Engineers.

Knaur, M., 1986, *Jasper National Park - Inventory of Contaminated Sites.* A paper prepared for Jasper National Park.

Leone, I.A., Flower, F.B., Gilman, E.F., and Arthur, J.J., 1979. Adapting Woody Species and Planting Techniques to Landfill Conditions. Environmental Technology Series, E.P.A. -600\2-79-128, U.S. Environmental Protection Ageny, Ohio, 122 p.

Lilley, S., 1985, *Resource Recycling in Alberta*, Published by the Environment Council of Alberta, Edmonton, Alberta.

Mason, Douglas A, and William G., 1988, *Statistics: An Introduction,* Harcourt Brace, Jovanocich Publishers, N.Y.

Messner and Assoc., 1974, *Measures of Effectiveness for Refuse Storage, Collection and Transport Practices*, Messner and Assoc., Silver Spring MD., Sponsor: National Research Centre, Cincinnati, Ohio, Solid and Hazardous Waste Research Lab.

National Contaminated Sites Remediation Program, National Classification System for Contaminated Sites, March, 1991.

National Guides for Landfills of Hazardous Wastes, 1991, Canadian Councils of Ministers of the Environment, Report CCME - WM/TRE - 02BE.

National Parks Act, 1980, National Parks Garbage Regulations, sor/80-217, Vol.114, p. 1091-1093.

National Parks Policy and Proceedures Manual, 1980, Part 8 of PRM, vol. 2, Published by authority of the Minister of the Environment , Minister of Supply and Services Canada, p. 40–41.

National Round Tables on the Environment and Ecomony, sponsored by Federation of Canadian Municipalities, Browning and Terris Industries, Ltd.

National Waste Reduction Handbook, 1991, National Round Table on the Environment and Economy, Federation of Canadian Municipalities, Sponsored by Browning - Ferris Industries Ltd.

Oakley, S., 1990, *A report prepared by the town of Banff Recycling Society*.

O'Brian, J. and J. William, 1991, *Calculating a Community's Maximum Recycling Potential*, HDR Engineering, Inc., Charlotte, North Carolina, Air and waste management.

O'leary, P. and P. Walsh, 1991, *Introduction to Solid Waste Landfills*, Waste Age, 1991.

O'leary, P., & B. Tansel, Oct. 1986, *Landfill Closures and Long Term Care*, Waste Age, p. 53–64.

Private Hauler Accepts California City Proposition, 1979 Waste Age v. 10(2), p. 17-25.

Rahn, T., 1987, *Garbage Incineration: Lessons from Europe & the United States,* A Report on the Polution Probe Foundation, Toronto, Ontario.

Rathye, W.L., 1991, *Once and Future Landfills*, National Geographic, v. 79(5), p. 116–125.

Recycling of Waste in Alberta, Environment Council of Alberta, 1987, Prepared by Environment Council of Alberta, Technical Report and Recommendation.

Reindle, J., 1977, *Interrelationships with the Solid Waste System*, Solid Waste Management Refuse Removal Journal, v. 20(4), p. 22-23, 54-55.

Robertson, R. and H. Bertrand, 1991, *Solid Waste: Issues and Answers*, from R. W. Beck and Assoc., Seattle, Washington.

Savage, G.M., L.P. Diaz, and C.G. Goluke, 1985, *Solid Waste Characterization,* Biocycle, v. 26(8), p. 35-37.

SCS Engineers, 1979, Long Beach, CA., *Development of Performance Specifications for Residential Refuse*, Sponsor: National Science Foundation, Washington, DC.

Siedel, C., 1992, *Recycling in Rural Communities: An Alberta Case Study*, A Masters Degree Project for the Faculty of Environmental Design, University of Calgary, Calgary, Alberta.

Siedel, C., 1993, *Town of Banff Recycling Program*, sonnevera Int. Corp.

Stanley and Associates, 1991, *Draft Waste Composition Study of Protocol for Regional Municipalities of Ottawa*.

Town Looks Ahead, 1978, Buys Bailer, Solid Waste Management Refuse Removal Journal, v. 21(8), p. 39-40.

Tulsa Refuse Department Competes for Routes, Employees and Landfill Space, 1979, Solid Waste Management Refuse Removal Journal, v. 22(5), p. 64-68, 89

Van Den Broek, E.and N.Y. Kiror, 1969, *The Characterization of Municipal Solid Waste*, Department of Fuel Technology, University of New South Wales, Australia.

Webb, C., 1983, *Use of Municipal Waste as Fuel*, Environment Council of Alberta, Edmonton, Alberta.

Woods, R., 1991, *Railhaul in the Long Run*, Waste Age, v. 22(12), p. 6-9.

Wright, J.R. & C. Neville, 1981, *Weigh Waste to Determine Quatity Before Recover Facility*, Solid Waste Refuse Removal and Liquid Waste Management, p. 32–34, 62–63.

Personal Communications

Bellingham, Gordon, Motor Transport Board, Feb 5, 1993.

Blanchett, R. Technical Officer, JNP, February 2, 1993.

Blascko, G., A/Fleet Supervisor JNP, November 10, 1992.

Boyle, C.,Assistant Instructor, Faculty of Environmental Design, 1992.

Breau, J., Information Officer, June, 1991.

Chambers, D., Townsite Manager, June, 1993.

Edwards, Dave, Chief of Highways and Scenic Corridors, JNP. Communication through Nov. 1992- Feb. 1993.

Gervais, J., Superintendent of Public Works, Banff, Alberta, 1993.

Greer, M. Highway and Sanitation Supervisor, JNP. January 25, 1993.

Goche, M. Local Operator for Grimshaw Trucking, Hinton Alberta, Feb. 5, 1993.

Herman, D., Supervisor of Waste Minimization Programs in Banff National Park, 1992.

Hieght, Paul, Registered Accountant for Bowen Trucking, Dec. 10, 1992.

Jones, S., Supervising Engineer with RW Beck & Associates, Chicago, Ill., 1993.

Koffin, F., Senior Citizens Paper Recycling, JNP. January 27, 1993.

Landry, A., President of the Jasper Recycling Society, Sept., 1992.

Lonsberry, D., Highways Roads Forman, JNP, December 10, 1992.

Marsh, J., P. Eng. Vice President, I.D. Systems Ltd., Calgary, Alberta, 1992.

Metcalf, K., Stanley & Associates (Centre), 1992.

OBerg, R., District Transport Engineer, Alberta Transportation, February 3, 1993.

Orydzuk,J., Plant Manager Paper Board Industries, January 15, 1993.

Pigeon, R., Sanitation Foreman, 1990.

Polski, R., Manager Hinton Regional Landfill, 1993.

Siedel, C., sonnevera International Corp. Environmental consultants, 1992.

Sime, Cal., Park Warden, Yoho National Park, 1993.

Stendie, Al, Sewage and Water Quality Officer, Jasper Park Warden Service, January 5, 1973.

Walker, A., Financial Project Manager, February 23, 1993.